William Daniel Dietz

The Soldier's First Aid Handbook

Comprising a series of lectures to members of the hospital corps and

company bearers

William Daniel Dietz

The Soldier's First Aid Handbook
Comprising a series of lectures to members of the hospital corps and company bearers

ISBN/EAN: 9783337309565

Printed in Europe, USA, Canada, Australia, Japan

Cover: Foto ©berggeist007 / pixelio.de

More available books at **www.hansebooks.com**

THE SOLDIER'S
FIRST AID HANDBOOK,

COMPRISING

A SERIES OF LECTURES

TO

MEMBERS OF THE HOSPITAL CORPS AND COMPANY BEARERS.

BY

WILLIAM D. DIETZ,

Captain and Assistant Surgeon, U. S. Army.

NEW YORK:
JOHN WILEY & SONS,
53 EAST TENTH STREET.
1891.

PREFACE.

THIS manual consists in the main of a series of lectures delivered to members of the Hospital Corps and company bearers, and covering the ground indicated in existing orders. Originality is not claimed for it; it is believed, however, that the subject-matter has been presented in available form. The writer's aim has been to supply an elementary "Soldier's Handbook," devoid of technicalities, and limited in its scope to the information required by the bearer who renders "First Aid," only. It is hoped, moreover, that the manual may contribute to facilitate the work of the medical officer in the preparation of his lectures to enlisted men, and, further, be of use to line officers who, in command of detachments, may have to meet emergencies in the absence of the surgeon.

ALCATRAZ ISLAND, CALIFORNIA, December, 1890.

TABLE OF CONTENTS.

	PAGE
PREFACE..	iii
PRELIMINARY REMARKS: The Organization and Duties of the Hospital Corps and Company Bearers.............	1

PART I.
THE HUMAN BODY.

THE SKELETON..	2
The Skull...	3
The Cranium..	3
The Face..	4
The Trunk..	5
The Spinal Column.................................	5
The Thorax...	5
The Pelvis..	6
The Limbs..	6
The Upper Limbs	6
The Lower Limbs...................................	7
The Joints..	8
THE SOFT PARTS..	9
The Muscles..	9
Fatty Tissue..	9
Connective Tissue..................................	9
The Skin..	9
The Organs of the Cranial and the Spinal Cavity.......	10
The Brain	10
The Spinal Cord....................................	11
The Cerebro-spinal Nerves........................	11

TABLE OF CONTENTS.

	PAGE
The Organs of the Lesser Cavities of the Head........	12
The Organs of the Chest............................	12
The Lungs 12, 13	
The Heart..................................13, 14	
The Organs of the Abdominal and Pelvic Cavities......	15
The Stomach..............................15, 16	
The Intestines............................	16
The Liver.................................	16
The Pancreas.............................	17
The Spleen...............................	17
The Kidneys..............................	17
The Ureters..............................	17
The Bladder..............................	17
The Urethra..............................	17
The Seminal Vesicles.....................	17
The Peritoneum..........................	18
Location of the Principal Blood-vessels...........18, 19, 20	

PART II.

FIRST AID ON THE BATTLE-FIELD.

	PAGE
GENERAL MANAGEMENT OF MEN WOUNDED IN BATTLE....	21
The Position of the Wounded upon the Litter........22, 23	
The Bearer's Equipment............................	24
The Use of First Aid Packets.....................25, 26	
FIRST AID TREATMENT OF HEMORRHAGE...............26, 27	
Capillary Hemorrhage.............................	27
Venous Hemorrhage...............................	28
Arterial Hemorrhage..........................28, 29	
Hemorrhage from the Artery of the Neck..........	30
" " " Artery of the Arm in its Uppermost Portion.......	30
" from the Artery of the Arm in its Lower Portion...........................	31
" from the Arteries of the Forearm.....	31
" " " " " " Hand........	31
" " " " " " Thigh........	32
" " " " " " Leg..........	32
" " " " " " Foot.........	32

	PAGE
Internal Hemorrhage	32
Fainting	33
FIRST AID TREATMENT OF WOUNDS	33, 34
Contusions	34
Contused Wounds	34
Incised Wounds	35
Punctured Wounds	35, 36
Lacerated Wounds	36
Gunshot Wounds	36, 37
Shock	37, 38
FIRST AID TREATMENT OF FRACTURES	38
Extemporary Splints	39
Fracture of the Upper Arm	39, 40
" " " Lower End of the Arm	40
" " " Forearm	40
" " " Finger	41
" " " Thigh	41
" " " Knee-pan	41
" " " Leg	41
" " " Spine	42
" " " Shoulder-blade, Hip-bone, or Rib	42
" " " Collar-bone	42
" " " Skull	42
" " " Jaw	42
Compound Fractures	42
FIRST AID TREATMENT OF DISLOCATIONS	43
FIRST AID TREATMENT OF SPRAINS	43
PROCEDURES TO BE ADOPTED IN CASES OF SUSPENDED ANIMATION	43, 44, 45
Artificial Respiration	44
Sylvester's Method	44, 45

PART III.

MANAGEMENT BY THE BEARER OF ORDINARY ACCIDENTS AND EMERGENCIES.

GENERAL RULES TO BE OBSERVED IN CASES OF ACCIDENT	46
HEMORRHAGE, WOUNDS, FRACTURES, DISLOCATIONS, SPRAINS. (*Vide* Part II.)	46

	PAGE
CONDITIONS CAUSING LOSS OF CONSCIOUSNESS	47
Fainting and Shock	48
Concussion of the Brain	48
Compression of the Brain	48
Apoplexy	48
Sunstroke	49
Heat Exhaustion	49
Intoxication	50
Epileptic Seizures	50
Poisons	50, 51
ASPHYXIA	51
Drowning	51, 52, 53
Strangulation from Hanging	53
Suffocation with Gases	53
Suffocation from Foreign Bodies in the Windpipe or Gullet	54
BURNS AND SCALDS	55
Slight Burns or Scalds	55
Burns or Scalds causing Blisters	55
Deep Burns or Scalds	56
Shock occurring as a Result	56
FREEZING	56
General Freezing	56
Frost-bites	56, 57
Chilblains	57
SORENESS OF THE FEET	57
HEMORRHAGE FROM THE NOSE, LUNGS, STOMACH, OR BOWELS	58
Nose-bleed	58
Hemorrhage from the Lungs	58, 59
" " " Stomach	59
" " " Bowels	59
POISONING	59–70
General Measures and Remedies	60, 61
Emetics	61
Alkaline Antidotes	62
Acid Antidotes	62
Bland Liquids	62
Stimulants	63
Laudanum	63

TABLE OF CONTENTS.

	PAGE
Forcible Administration of Remedies	63, 64
Poisoning from Unknown Substances	64
Corrosive Poisons	64
Sulphuric, nitric, and hydrochloric acids	65
Carbolic acid	65
Oxalic acid	65
Ammonia, soda, and potash	65
Corrosive sublimate	65, 66
Nitrate of silver	66
Phosphorus	66
Irritant Poisons	66
Arsenic	66, 67
Tartar emetic	67
Lead	67
Copper	68
Iodine	68
Irritant animal and vegetable substances	68
Tainted meat, tainted fish, toadstools	68
Neurotic Poisons	68
Opium	69
Chloral	69
Hydrocyanic acid	69
Irritant narcotics	70
Poisoned Wounds	70
Bites of venomous serpents	70
Bite of a mad dog	70, 71
Stings of tarantulas, etc	71
Rhus Poisoning	71
TABLE OF POISONS	71, 72, 73, 74
FOREIGN BODIES IN THE EYE, EAR, OR NOSE	74, 75, 76
Foreign Bodies in the Eye	74, 75
Foreign Bodies in the Ear	76
Foreign Bodies in the Nose	77
MISCELLANEOUS HINTS	77, 78
Constipation	77
Colic	77, 78
Cholera Morbus	78
Diarrhœa	78
SIGNS OF DEATH	78, 79
BOOKS OF REFERENCE	81

THE SOLDIER'S FIRST AID HANDBOOK.

PRELIMINARY REMARKS.

Organization and Duties of the Hospital Corps and the Company Bearers.

According to the regulations governing the United States Army, the Hospital Corps consists of hospital stewards, acting hospital stewards, and privates regularly enlisted for and attached to the Medical Department, and performing all hospital services in garrison and in the field. In time of war, this corps renders the necessary ambulance service under such officers of the Medical Department and assistants as may be duly detailed.

Moreover, the regulations require that in each company throughout the army four men, to be known as *company bearers*, be designated for instruction in the duties of litter-men and the methods of rendering first aid to the disabled. During an engagement, acting under orders of their commanding officers and supervision of their regimental surgeon, they render first aid to their wounded comrades and carry them to the rear. Upon being relieved by members of the Hospital Corps, they immediately join their companies.

Collectively, privates of the Hospital Corps and the men designated from the companies will herein be known, for the sake of brevity, as *bearers*.

PART I.

THE HUMAN BODY.

In order that the bearer may be enabled intelligently to afford first aid to the injured or to care for the sick, he requires, first of all, a general knowledge of the structure of the human body and of the functions of the principal organs. Armed with this knowledge, elementary though it must be, he will frequently be in a position to give untold relief to his stricken comrades, or even, by the application of a few simple principles learned, to save a life that without his intervention would have been lost.

The human body is composed of the *skeleton* and the *soft parts*.

The **Skeleton** comprises the *bony parts* of the system. It is the framework of the body, supports the soft parts, encloses the vital organs in its cavities, and furnishes a system of levers acted upon by the muscles. Bones may be classed as *long, short, flat,* or *irregular,* according to their shape. The thigh-bone, for instance, is a long bone; the bones of the wrist are of the short variety; the hip-bone is a flat bone, and the lower jaw an irregular one. The skeleton may be divided into *skull, trunk,* and *limbs;* in the adult it is composed of two hundred and thirty-two bones, including the teeth.

The **Skull** consists of the *cranium,* which contains the brain, and the *face.*

The **Cranium** is made up of the eight bones which enclose the cranial cavity. The cranial bones are:

One *occipital bone,* which is located at the back and lower portion of the skull, and through a large opening in which the brain and spinal cord communicate;

Two *parietal bones,* forming by their union the sides and roof of the skull;

One *frontal bone,* whose vertical portion forms the forehead, while its horizontal portion enters into the formation of the roof of the cavities of the eyes (the orbits) and of the nose;

Two *temporal bones,* forming the temples and containing the inner ear;

One *sphenoid bone,* situated at the base of the skull, and joining with all the other bones of the cranium and some of the face; and

One *ethmoid bone,* situated at the root of the nose, and entering into the formation of the cavities of the skull, the eyes, and the nose.

The cranial bones are joined to their neighbors

by means of dovetailed, toothed, or bevelled edges. The lines of junction are known as *sutures* (seams).

There are a number of openings in the cranial bones through which blood-vessels and nervous structures pass into or out of the cranial cavity.

The **Face** is formed of fourteen bones, viz.:

Two *nasal bones*, forming the bridge of the nose;

Two *superior maxillary bones* (upper jaw-bones);

One *inferior maxillary bone* (lower jaw-bone);

Two *malar* (cheek) *bones;*

Two *palate bones;*

One *vomer*, forming a portion of the partition of the nose;

Two *lachrymal bones*, forming part of the inner wall of the orbit or cavity of the eye; and

Two *turbinated bones*, situated in the nasal cavities.

Situated above the Adam's apple, and connected by muscles to some of the bones of the skull, is the *hyoid bone*, which may well be mentioned with them.

The bones of the face enter into the formation of the cavities of the eye, nose, and mouth.

The *teeth* are sometimes classed as bones. They are thirty-two in number in the adult, viz., four *cutters* or *incisors*, two *canine* (so called on account of their resemblance to a dog's teeth), and ten *grinders* (four *false molars*, six *true molars*) in each jaw.

THE TRUNK.

The **Trunk** comprises the *spinal column*, the *thorax*, and the *pelvis*.

The **Spinal Column**, or *back-bone*, consists of twenty-four bones, known as *vertebræ*, each of which is formed of a *body* and of certain *offshoots* from that body. The body and its offshoots enclose a round opening. In the natural arrangement in column of the vertebræ, these openings form a long canal, the *spinal canal*, which contains the spinal cord, and communicates with the cranial cavity, or cavity of the skull, through the large opening in the occipital bone previously referred to.

The **vertebræ** are connected by tendon-like substances permitting of a bending and a twisting of the back-bone. They are divided into—

Seven *cervical*, situated in the neck;

Twelve *dorsal*, situated at the back of the chest;

Five *lumbar*, situated in the region of the loins.

The cervical vertebræ are the smallest, the lumbar the largest. The vertebræ in general resemble one another in shape, with the exception of the upper two cervical, which are modified to form joints for movements of the head.

The **Thorax**, or *chest*, is composed of the *sternum*, or *breast-bone*, and the *ribs*.

The *sternum* is a flat, narrow bone, situated in the middle line in front of the chest. In shape it somewhat resembles the old Roman sword; it is broad above and pointed below.

The *ribs* are twenty-four in number, twelve on each side. They are elastic arches of bone, and are joined behind to the dorsal vertebræ. In front seven pairs are joined to the sternum by means of the *costal cartilages;* these are the *true*, while the other five pair are the *false ribs.* In front the upper three pair of false ribs are attached by their cartilages to the cartilages of the true ribs ; the lower two pair having the front ends free are called *floating ribs. Cartilage* is popularly known as *gristle.*

The **Pelvis** (*pelvis*, a basin) forms the lowest portion of the trunk, and is composed of the two *hip-bones*, the *sacrum*, and the *coccyx.* The *hip-bones* are flat bones lying in front and on the sides of the pelvis. On the outer side of each hip-bone is a cup-shaped depression for the head of the thigh-bone. The *sacrum* is situated at the back of the pelvis, and is joined to the last vertebræ and the hip-bones. The *coccyx* joins the lower portion of the sacrum.

The *pelvic cavity* is continuous with the *abdominal*, which is formed by the five lumbar vertebræ, and certain soft parts.

THE LIMBS.

The **Limbs** are divided into an *upper* and a *lower pair*.

The **Upper Limb** consists of the *shoulder, arm, forearm*, and *hand*.

The **Shoulder** consists of the *clavicle*, or *collar-bone*, and the *scapula*, or *shoulder-blade*. The former is a bone shaped somewhat like the italic letter *f*, and joins with the sternum and the scapula. The scapula is a flat, triangular bone, which articulates (joins) with the humerus to form the shoulder-joint.

The **arm** consists of one long bone, the *humerus*.

The **fore-arm** consists of two long bones, the *ulna* and the *radius*, the former situated on the little finger side of the arm and entering into the formation of the elbow-joint, the latter situated on the thumb side and forming a large part of the wrist-joint.

The **hand** is composed of the *carpus*, or *wrist* (formed of eight short bones), the *metacarpus*, or part between the wrist and the fingers (five small *long* bones), and the *fingers* (fourteen small *long* bones, three to each finger, with the exception of the thumb, which has two).

The **Lower Limb** consists of the *thigh*, the *leg*, and the *foot*.

The **thigh** consists of one long bone, the *femur*, or *thigh-bone*, whose upper end, called the *head*, fits into the cup-shaped receptacle of the hip-bone.

The **leg** consists of two long bones: the inner and stronger one is called the *tibia* (shin-bone), and the outer and weaker one the *fibula*. Their lower ends form the *ankles*.

Between the thigh and leg, in front, is situated the *patella*, or *knee-pan*, a round, flat bone.

The **foot** consists of the *tarsus*, which corresponds to the *wrist*, the *metatarsus*, and the *toes*. The *tarsus* consists of seven short bones, the largest of which is the *os-calcis* (heel-bone), to which the prominent tendon of the back leg is attached. The *metatarsus* and the *toes* are similar in structure to the *metacarpus* and the *fingers*, and contain five and fourteen bones, respectively.

THE JOINTS.

The various bones of the body are held together at different parts of their surfaces, the connections being known as *joints* or *articulations*. Joints may be classed as *immovable, slightly movable,* and *freely movable.* The joints of the cranial bones, for example, are immovable, those of the spine in general, slightly movable, the knee and elbow are freely movable. The parts forming a freely movable joint are the *ends of the bones*, the *cartilages*, or gristly substances, covering these ends, the *ligaments*, or bands which hold them together and form a *capsule* or *sac* around the joint, and a *synovial membrane* lining the interior of the joint and secreting a thick lubricating fluid. Among the subdivisions of the movable joint are the *gliding-joint*, as the one between breast-bone and collar-bone, the *hinge-joint*, as exemplified in the elbow and knee, and the *ball* and *socket joint*, of which the shoulder-joint and hip are types.

THE SOFT PARTS.

The **Muscles** constitute the bulk of the soft portions of the body. Their function is by contracting to move the parts into which they are inserted. They may be classed as *voluntary* or *involuntary*, according as they are or are not under control of the will. The muscles of the arm are examples of the voluntary, those of the bowels of involuntary, muscles. Contraction is brought about by an impulse originating in the brain and communicated to the muscle by means of a nerve. Voluntary muscles, as a rule, originate from and are inserted into bone. The *origin* is the *more fixed end*, the *insertion* the *movable point* to which the muscular power is applied. The origin, however, is absolutely fixed only in a few muscles, such as those of the face, which originate from bone and are inserted into movable skin; in the greater number of cases the muscle may be made to act *from either end.*

Fatty Tissue constitutes another of the soft parts. It is widely distributed, imparting a rotund fulness to the form, preventing a too rapid dissipation of bodily heat, protecting the internal organs against cold, and lessening the effects of shock and pressure upon the external parts.

Connective Tissue is distributed throughout the body for the purpose of holding in position the component structures.

The **Skin** is the outermost covering of the

body. It is the principal seat of the sense of touch, and serves as a protection for the deeper tissues. It is composed of an outer layer, the *scarf-skin*, and an inner layer, the *true skin*. The *nails* are a modification of the scarf-skin. Connected with the true skin are the *sweat-glands*, the *hair-follicles*, and *glands* secreting an *oily* material. Imbedded in the true skin are innumerable small *blood-vessels* and fine *nerves*. The skin is very absorbent.

Among the remaining soft parts are the blood-vessels, the lymphatic vessels which carry nutritive fluid from the intestines and elsewhere into the circulation, the nerves, and a number of organs which will be considered hereafter.

ORGANS OF THE CRANIAL AND THE SPINAL CAVITY.

The *cavity of the skull* contains the *brain*, the *membranes* covering it, and the *nerves* connected with it. The lower and posterior part of the brain joins with the spinal cord through the large opening in the base of the skull. The nerves arising directly from the brain are called *cranial* nerves, and are twelve in number. They are principally distributed to the region of the head. The *spinal canal* contains the *spinal cord*, its *membranes*, and the thirty-one pairs of so-called *spinal nerves* given off at regular intervals at each side of the column.

The **Brain** is endowed with a number of faculties, among which may be mentioned the faculty of perception, the intelligence, and the will. In the brain originate the impulses governing voluntary motion. This organ also exercises control over many of the bodily functions which need not be specified here.

The **Spinal Cord** is a continuation of the brain. It sends nerves to the muscles and integuments (coverings) of the trunk and limbs. It acts as an organ of communication between the brain and the external parts, but has special functions of its own in addition.

The brain and spinal cord together are known as the *cerebro-spinal axis*, and a nerve originating from this axis is called a *cerebro-spinal nerve*. There are the nerves concerned with nutrition, growth, etc., known as *sympathetic nerves*, whose consideration, however, does not come within the scope of this manual.

Cerebro-Spinal Nerves are a series of rounded cords arising from the cerebro-spinal axis, and distributed to all portions of the body. As before mentioned, those arising *directly from the brain* are called *cranial nerves*, those *arising from the spinal cord, spinal nerves*. It must be remembered, however, that even the latter are ultimately connected with the brain.

Nerves may be divided into *sensory* and *motor*, or *nerves of sensation* and *nerves of motion*. The former convey sensations to the brain, the latter carry from brain to muscle the impulses under which muscular contraction is produced.

The *cranial nerves* include those of *smell*, *sight*, *hearing*, *taste*, one nerve of *touch*, and several *motor* nerves.

The *spinal nerves* give off both *sensory* and *motor* branches.

THE ORGANS OF THE LESSER CAVITIES OF THE HEAD.

The *lesser cavities* of the head are the *orbits* containing the eyes, the *nasal cavities*, containing the apparatus of smell, the *cavity of the mouth*, containing the organ of taste, and the *cavity of the internal ear*.

THE ORGANS OF THE THORACIC CAVITY.

The *thoracic cavity* or *chest* is one of the large cavities of the trunk, the other two being the *abdominal* and the *pelvic cavity*. The chest is separated from the belly by a muscular partition called the *diaphragm* (midriff).

The *heart* and *lungs* occupy the larger portion of the chest, which further contains a part of the *wind-pipe* and *gullet*, the *large blood-vessels*, and the *pleuræ*.

The **Lungs** are divided into a *right* and *left* lung. The right has three lobes, the left two. One layer of a membraneous sac called the *pleura* covers each lung, the other layer lines the adjacent portion of the interior of the chest.

The lungs are the main organs of respiration or breathing. An adult breathes ordinarily about *twenty times a minute*. On inspiration, fresh air is drawn into the lungs, and is taken up by countless numbers of minute blood-vessels distributed through their substance, the blood thus obtaining the pure oxygen it needs. On expiration the impure gases which the blood gives off are expelled. Air reaches the lungs through the *mouth* or *nose*, the *larynx*, which projects in front of the neck as the *Adam's apple*, the *trachea*, or *windpipe*, and the *bronchial tubes*. The *main bronchial tubes* are two in number, one for each lung. The *right tube* subdivides into three, the *left* one into two branches, to supply the five lobes. This subdivision continues until the smallest tubes are reduced in diameter to from one-fiftieth to one-thirtieth of an inch.

The **Heart** lies behind the lower two-thirds of the sternum, or breast-bone, and projects farther into the left than into the right side of the chest. It is five inches in length, three and one-half inches in width at the broadest part, and two and one-half inches in thickness. The widest portion is uppermost; the point, or *apex*, is downward, and may be felt beating between the fifth and sixth rib on the left side. By a vertical partition the heart is divided into halves, called respectively the right and the left; a horizontal partition divides each half into two cavities, the upper ones called the *auricles*, the lower the *ventricles*. The auricle of each side opens into the ventricle of the same side, the opening being guarded by a valve.

From the left ventricle arises the main artery of the body, the *aorta*. By an *artery* is meant a vessel which carries blood *away* from the heart, by a *vein* one which carries blood *towards* it. *Capillaries* are the minute vessels intermediate between the smallest arteries and the smallest veins. The aorta gives off branches which divide and subdivide until the capillaries are reached. These unite to form small veins, which joining together, form larger and larger vessels, these finally opening into the right auricle.

The heart is mainly composed of muscle, and may be regarded as a muscular force-pump. When it contracts, it forces blood from the ventricle successively into the aorta, the lesser arteries, the capillaries, the veins, and into the right auricle. This course of the blood constitutes the **systemic circulation,** or the circulation through the body at large.

The blood which left the left heart pure has in its course taken up noxious gases, which must be given off in the lungs in exchange for the oxygen of the air. Hence the blood reaching the right auricle passes into the right ventricle, and is pumped through the lungs by a contraction of the right ventricle. In the lungs it courses through the *pulmonary arteries*, the *pulmonary capillaries*, and the *pulmonary veins*, reaching the left auricle and finally the left ventricle in a purified condition. This is the **pulmonary circulation** or circulation through the lungs.

When an artery of the systematic circulation is cut, the blood issues in spirts, and is of a bright-red color; venous blood flows in a steady stream and is dark red; in capillary hemorrhage the blood oozes out, and is of an intermediate shade.

When the finger is laid upon a superficial artery a series of shocks is perceived known as the *pulse*, and due to the contractions of the heart acting upon the column of blood. The pulse-rate ordinarily in the adult is seventy; that is to say, his heart beats that number of times in a minute.

As connected in a measure with the subject of circulation, it is to be noted that the *normal temperature of the human body* as registered under the tongue is ninety-eight and nine-tenths degrees (98.9°) F.

THE ORGANS OF THE ABDOMINAL AND PELVIC CAVITIES.

The *abdominal cavity* contains the following organs: the *stomach*, the *intestines* or *bowels*, the *liver*, the *pancreas*, the *spleen*, the *kidneys*, and the *peritoneum*, or membrane covering the bowels. It is separated from the cavity of the thorax by the *diaphragm*. The *pelvic cavity* is continuous with the abdominal, and contains the *bladder* and its *appendages*, a portion of the *generative apparatus*, and the *lower end of the bowels*.

The **Stomach** is the principal organ of digestion. It resembles a sack in shape, and is situated immediately behind the anterior wall of the abdomen below the liver and the diaphragm. When moderately full, its horizontal diameter measures about twelve inches, its vertical about four. It has two openings, one on the left side for the *gullet*, or food passage connecting the stomach with the mouth, and one on the right for the bowel. The stomach secretes *gastric juice*, which is one of the digestive fluids.

The **Intestines** are divided into the *small* and *large intestine*. The *small intestine*, about twenty feet in length, begins at the stomach and ends in the large intestine. The *large intestine* is about five feet in length, and extends from the small intestine to the external opening of the bowels, the *anus*. It lowest portion is called the *rectum*.

The **Liver** lies mainly upon the right side of the abdominal cavity under the diaphragm. It is composed of two lobes,—a larger, the *right lobe*, and a smaller, the *left lobe*. The main function of the liver is to extract from the blood the substances forming bile; besides this, it also brings about certain changes in the constituents of the circulatory fluid. The biliary substances are of two classes, viz., such as are of further use in the system, namely, in the process of digestion, and those which have been extracted from the blood as impurities, and are hence to be discharged from the body as valueless. Bile is stored in the *gall-bladder*, from which it is conducted into the intestine through a tube, the *bile-duct*.

The **Pancreas** lies behind the stomach. It secretes a fluid which assists in the process of digestion.

Organs which, like the liver and pancreas, have the faculty of extracting certain substances from the blood and elaborating them into fluids destined for various purposes, are called *glands*. The fluids formed are called *secretions*, if they are to be further utilized in the body, and *excretions*, if they are impurities and destined to be cast off.

The **Spleen** lies on the left side of the abdominal cavity, opposite the liver. Its functions are concerned in the elaborations of the blood.

The **Kidneys** are the two glands lying one on each side of the spinal column. They are about four inches in length, and bean-shaped. Their function is to separate from the blood certain substances, which, when dissolved in water likewise withdrawn from the circulatory fluid, constitute the *urine*. The daily quantity of this excretion is from forty to sixty ounces (about a quart and a quarter to two quarts).

The **Ureters** are the two tubes which conduct the urine from the kidneys to the bladder.

The **Bladder** is situated in the pelvic cavity and is a receptacle for the urine.

The **Urethra** is the tube which discharges the urine from the bladder. For a part of its course it runs through the male organ, the *penis*.

The **Seminal Vesicles,** or sacs containing the seminal fluid, lie underneath the bladder. They store the semen formed in the testicles, discharging it through small ducts into the urethra.

The **Peritoneum** is a membranous sac one layer of which covers the abdominal organs and some of the pelvic, while the other layer lines the interior of the walls of the belly. It forms a number of *ligaments*, or bands, which hold the organs in place, and secretes a lubricating fluid called *serum*.

The abdominal and pelvic cavities together are known popularly as the *belly*.

LOCATION OF THE PRINCIPAL BLOOD-VESSELS.

One of the most important anatomical points with which it is necessary that the bearer should be familiar is the location of the large arteries. Especially should he know where the vessels are superficial, that is to say, almost immediately under the skin.

The **Aorta** is the main vessel of the systemic circulation. It arises from the left ventricle, ascends for a short distance behind the breast-bone, then forming its *arch*, reaches the spinal column, and descends within the chest and abdomen, on the left side of the spine, dividing at the level of the fourth lumbar vertebra. The arch gives off arteries whose branches supply the head and upper limbs, while the descending portion furnishes vessels to the trunk and its organs, and on dividing gives off a large artery to each of the lower limbs.

The **arterial trunks supplying the head** ascend, one on each side of the windpipe, and are superficial along the edge of the diagonal muscles, which are prominently seen when the head is turned to one side or the other. They are known as the *carotids*.

The **arterial trunk supplying the upper limb** emerges from the chest over the first rib, and passing under the middle of the collar-bone, runs through the axilla or arm-pit, and along the inner side of the arm between the two prominent muscles, the biceps and the triceps. At the middle of the front of the elbow it divides into two principal branches for the fore-arm. It is superficial in the hollow behind the collar-bone, at the apex and outer side of the arm-pit, in the groove between the biceps and triceps, and in front of the elbow.

The two **principal branches for the fore-arm** pass down, one, the *radial*, on the thumb (outer) side, and the other, the *ulnar*, on the little finger (inner) side of the limb. In the palm they form a loop, the *palmar arch*, which gives off branches to the fingers. The *radial* is superficial throughout its entire extent, and may be felt beating at the outer side of the outer of the two tendons, which are prominent in front of the middle of the wrist. The radial artery at the wrist is generally felt for the *pulse*. The *ulnar* is deeply situated for the greater part of its course, but at the wrist it is superficial, and located about half an inch to the inner side of the inner of the two prominent tendons mentioned.

The **arterial trunk supplying the lower limb** arises from the termination of the aorta near the end of the back-bone, and passes out through the groin to the front of the thigh at about two-thirds the distance from the hip-bone to the middle line of the body. Descending between the muscles on the inner side of the thigh, it reaches the middle of the hollow at the back of the knee, subsequently dividing into three principal branches for the leg. It is superficial for several inches from the groin down, and again in the middle of the hollow of the knee.

The **principal arterial branches for the leg** are three. One passes forward through the space between the tibia and fibula, and descends between the muscles of the leg, becoming superficial in front of the middle of the ankle joint. The other two branches are deeply situated in the back of the leg, one on each side, becoming superficial behind the inner and outer ankle, respectively. They supply the sides and sole of the foot, and in the sole unite in a loop, the *plantar arch*, from which branches are distributed to the toes.

The **veins** generally follow the course of the arteries. There are a number of them, however, which are not, so to speak, mated with arteries.

PART II.

FIRST AID ON THE BATTLE-FIELD.

Besides carrying out instructions concerning cases which the surgeon has already treated, the bearer, independently, will frequently be called upon to administer *first aid* to his disabled comrades. Especially is this the case in time of war, and hence it is meet that the management of those wounded in battle should be first considered.

GENERAL MANAGEMENT OF MEN WOUNDED IN BATTLE.

In the absence of anything demanding prior attention, the bearers on reaching a wounded comrade should first of all contribute to his comfort by opening tight garments and removing encumbering accoutrements. The rifle or carbine should be unloaded, not discharged. While one or two bearers are accomplishing these things, a third administers water or necessary stimulant, while a fourth busies himself with examining the wound. In order to obtain a view of the injury, it may be necessary to cut the investing clothes along the seam, or to remove the garment. When the latter is done, it is to be remembered that the injured part is the last to be unclothed, while it is the first to be clothed when the garment is replaced. The wound having received proper attention, the severely injured patient is carried from the field to the so-called *dressing station*, distinguishable by the red-cross guidon or a lantern. Here a complete examination is made by the surgeon, food is given, and a classification of cases effected, prior to their transfer, if necessary, to the field hospital.

POSITION OF THE WOUNDED UPON THE LITTER.

When a patient is carried upon a litter he should be made as comfortable as possible, strained and painful positions being avoided. For the purpose of bolstering him up, pillows, blankets, various available articles of clothing, knapsacks, bundles of hay, etc., may be used. Fractured limbs must be securely steadied, either by the application of temporary splints, etc., or by being appropriately propped up. *In hemorrhage from the great cavities the body should be inclined towards the bleeding wound. Unconsciousness resulting from faintness requires that the head should be kept low, but when it is due to a blow upon the skull, the head should be raised.* An extemporary pillow with side cushions may be made by folding a blanket or overcoat to the desired width, rolling it from both ends, and after proper adjustment, securing it in position by a bandage, handkerchief, or neck-tie. It is especially useful in wounds of the head and face. *Gaping wounds of the front of the neck require that the head be well raised and bent forward upon the chest. Penetrating wounds involving the lung frequently cause great distress in breathing, which the bearers must seek to remedy by tucking a folded blanket, overcoat, blouse, under the raised chest.* In penetrating wounds of the abdomen, in general, the patient is placed upon his back if the wound is in front, and upon the injured side if lateral. In both cases the legs should be drawn up, and in the first instance the shoulders raised. A horizontal slash across the belly requires the sitting posture, a vertical one the recumbent. In wounds of the upper arm the patient is laid upon his back, the injured extremity evenly and comfortably supported, and resting alongside the body. In injuries of the fore-arm, wrist, or hand the wounded member is best laid across the chest or abdomen, with the elbow propped up, and the arm, if practicable, bound to the body. In injuries of the lower extremities the patient is laid upon his back and the member securely propped; in case of fracture of one limb, the other limb should be utilized as a splint, the wounded one being tied to it. Improvised splints may be applied if the patient is to be transported for a long distance.

LIMITATION OF THE DUTIES OF BEARERS.

The bearers should be careful never to exceed their duties by endeavoring to afford treatment that had better be applied by the surgeon, but should strive to place the wounded man under that officer's care as soon as practicable.

THE BEARER'S EQUIPMENT.

Pursuant to regulations, each member of the Hospital Corps and each company bearer in the field or in time of war, carries upon his person a canteen of water, a knife of approved pattern, and a simple package of dressings; one-fourth of the hospital privates carry the so-called *medicine-cases*, containing portable drugs, dressings, restoratives, anæsthetics, and a few simple instruments.

The package of dressings carried by all bearers, and one similar to which is in certain foreign armies issued to every soldier, is known as "*Esmarch's First Help for Wounds.*" It is a flat packet about four inches by three in dimension, and contains, in a water-proof wrapper, two antiseptic compresses of sublimated gauze in oiled paper, one antiseptic bandage of sublimated cambric, one Esmarch's triangular bandage with its mode of application illustrated upon itself, and two safety-pins. Printed directions upon the wrapper read as follows:

THE USE OF FIRST AID PACKETS. 25

"Place one of the compresses on the wound, removing the oiled paper. In case of large wounds open the compress and cover the whole wound. Apply the antiseptic bandage over the compress. Then use the triangular bandage as shown by illustrations on the same."

In order that he may be able to properly dress such injuries as he is called upon to attend, the bearer must be provided with appliances to check hemorrhage, exert pressure upon parts, retain them in their proper position, or protect them against dirt. Such articles include lint, compresses, roller and triangular bandages, splints, and either tourniquets or elastic bandages.

Lint is shredded or scraped linen used to cover wounds and to arrest slight bleeding. It is either applied dry, or moistened with water, oil, vinegar, antiseptic fluids, etc. *Patent lint* is a manufactured substitute furnished in sheets. Allied to lint are *styptic cotton*, which coagulates or clots blood, *absorbent cotton*, *antiseptic cotton*, etc.

A *compress* is a piece of linen or muslin folded upon itself several times, retaining the lint in place and serving as a pad to exert pressure.

Roller bandages consist of rolled strips of certain materials of various lengths and breadths, and are applied over the so-called first pieces of surgical dressing which are in immediate contact with the wound. They are generally made of linen, flannel, or gauze. A roller bandage should be applied with regularity, so that the pressure exerted may be uniform. It should be neither too tight nor too loose, and when employed upon an extremity, should be made to ascend the limb.

The *triangular bandage* is not only useful as a retentive dressing, but also as a sling. Furthermore, it may be employed in the form of the knotted cloth, or of the Spanish windlass, to be described hereafter.

Splints are made of wood, paste-board, leather, wire, felt, straw, or other material, and are used to keep fractured limbs in a fixed position. They may be improvised from a number of articles available on the battle-field, such as rifles, carbines, scabbards, side-arms, rolled blankets, pickets, shingles, sticks, twigs, straw mats, etc. Improvised splints should be well padded to conform with the curves of the limb. Splints are kept in place by triangular or roller bandages, straps, cords, etc. Under *Fractures*, they will be referred to more fully.

The *tourniquet* and the *elastic bandage* are used to exert pressure upon blood-vessels. They will be further considered under hemorrhages.

It will be noticed that the *Esmarch's package* contains articles subserving the purposes of all the above appliances, splints excepted.

FIRST AID TREATMENT OF HEMORRHAGE.

Upon the field of battle the casualties claiming the attention of the bearer will mainly be wounds. Hemorrhage or bleeding always accompanies these injuries, and as a necessary preliminary to an intelligent management of them a knowledge of its first aid treatment is essential. Serious hemorrhage is an accident, in the treatment of which the application or omission of a few simple procedures, easily learned, easily remembered, and readily applied, will often make the difference of life and death to the sufferer.

The mechanism of the circulation and the location of the principal arteries have already been considered.

Hemorrhage, as stated, is of three kinds, viz., in the order of gravity, capillary, venous, and arterial.

CAPILLARY HEMORRHAGE.

Capillary Hemorrhage, alone, attends every minor cut. The blood oozes out and is of a color intermediate in shade between that of arterial and venous blood. The flow is generally inconsiderable, and ordinarily is soon arrested by the exposure of the cut to the air. Elevation of the parts and pressure upon them may effect the same result, or cold, hot, astringent, or styptic (blood-clotting) applications may be employed. Vinegar, alum, and tannin are astringents; Monsel's solution is a styptic. Hot applications, such as a folded cloth soaked in as hot water as the hand will tolerate and lightly wrung out, are of especial use in oozing from large surface.

VENOUS HEMORRHAGE.

Venous Hemorrhage occurs as a steady flow, the blood being of a dark shade of red. Generally it is readily controlled—when slight, even by the simple application of cold, or when severer, by the employment of pressure above and below the wound in the vessel. This pressure may be made by applying the fingers or a dry pad of cloth firmly bound down by a bandage, or in some cases both.

ARTERIAL HEMORRHAGE.

Arterial Hemorrhage occurs in jets, their number corresponding to the number of heart-beats in the same time. The blood is bright red. When the largest arteries are wounded, the resulting hemorrhage, if unchecked, soon causes death on account of the great amount of blood lost. In injuries of this kind, all that the bearer will be able to do will be to thrust his finger or some other plug deeply into the wound and to endeavor by firm pressure to arrest the flow. It is rarely that he will succeed in his purpose.

As arteries carry the blood away from the heart, the general rule to be observed in the management of arterial hemorrhage is to *completely obstruct the artery by pressure at the bleeding point or between it and the heart.*

Pressure may be applied by the fingers, pad or plug and bandage, knotted cloth or handkerchief,

TREATMENT OF ARTERIAL HEMORRHAGE. 29

Spanish windlass, tourniquet, the elastic bandage, or by flexion (bending) of an extremity so that the vessel is compressed between the members.

The *finger* is applicable in all cases; the *pad and bandage* may be employed in bleeding from quite small branches. The *plug* consists of a packing of lint, cotton (styptic or otherwise), old muslin or linen, etc., thrust into a wound and firmly bound down by a bandage. In using the *knotted handkerchief*, the knot is placed at the proper spot over the course of the artery, and the two ends of the handkerchief are then passed around the limb and firmly tied. The Spanish windlass and the tourniquet are modifications of the knotted handkerchief. The *windlass* consists of a cloth into which a stone or other hard round body has been folded. The stone being placed at the proper spot over the course of the artery, the ends of the cloth are brought around the limb and loosely tied. A stick is slipped under and twisted, until the tightened cloth, through the stone in it, causes *just enough* pressure upon the vessel to stop the bleeding. In order to avoid pinching of the skin under the stick, a pad of some kind may, if available, be interposed. The *field tourniquet* consists of a pad, which takes the place of the knot in the handkerchief, and a strap and buckle to hold the pad in its position over the artery. The *screw tourniquet*, in addition, is furnished with two plates, through one of which works a screw. When the instrument has been adjusted the turning of the screw tightens the strap and increases the pressure exerted by the pad upon the artery. Sometimes the pad is wanting, when its place may be filled by a roller. The *elastic bandage* is a roll of rubber band about three inches wide. It causes considerable pressure by virtue of its elasticity. It is applied at a point immediately above the wound, and is made to spirally ascend the member. The influence of *flexion of a limb* upon the arterial flow in the vessels below the joint can be seen by noting the cessation of the radial pulse when the forearm is forcibly bent upon the arm. Flexion may be combined with the use of the knotted cloth.

In cases of hemorrhage controlled by the tourniquet, windlass, elastic bandage, or knotted handkerchief, the surgeon's services should be procured as soon as practicable, otherwise the continuous pressure exerted will work injury by shutting off the circulation from the limb below.

For **Hemorrhage from the Artery of the Neck,** pressure with the fingers is alone indicated. It is applied by pushing the finger deeply into the neck in a backward and inward direction at the anterior border of the prominent neck muscle.

For **Hemorrhage from the Artery of the Arm in its Uppermost Portion,** downward pressure is applied behind the middle of the collar-bone. Naturally this pressure will first be made by the thumb, which may be moved towards the breastbone or towards the shoulder if it fails to strike the artery at once. If necessary, the thumb may be cautiously replaced by a finger-shaped, nicely rounded stone or stick, or a rifle cartridge, etc., the skin being protected as soon as practicable by an interposed pad of lint, linen, or bandage.

For **Hemorrhage from the Artery of the Arm in its Lower Portion,** pressure by the thumb is made above the wound and upon the vessel's course (i.e., along the inner side of the arm, and in the line of division between the two prominent muscles), the arm being grasped between the thumb and fingers of the bearer's hand. Subsequently a piece of cloth tied into a knot as big as a fist may be pushed well up into the armpit, and the arm brought down and bound against the side of the chest; or the Spanish windlass, tourniquet, or elastic bandage may be applied.

For **Hemorrhage from Arteries of the Forearm,** pressure by the fingers on the artery of the arm, made as above detailed, is the first measure to be adopted. A knotted handkerchief, the knot placed over the middle of the joint, is then tightly tied around the limb, and the fore-arm bent so as to press forcibly against the knot. The main blood-vessel situated in the middle line over the elbow will then become occluded. Instead of the knotted handkerchief, the elastic bandage may be used. It is applied above the wound, and its successive turns are made to ascend the arm.

For **Hemorrhage from the Arteries of the Hand,** pressure is made with the fingers upon the bleeding spot, or with both thumbs on the arteries on each side of the prominent wrist tendons, together with elevation of the part. Again the bandage and pad may be employed, or, if necessary, the measures detailed in the preceding paragraph.

For **Hemorrhage from the Arteries of the Thigh,** pressure by the thumbs, the fingers grasping each side of the limb, is the first measure to adopt. This pressure is made over the course of the artery and above the wound. The vessel makes its appearance in the groin at about two-thirds the distance from the hip-bone to the central line, and is superficial for several inches below this. Pressure may further be made by means of a padded stick, a Spanish windlass, a tourniquet, the elastic bandage, or by the knotted handkerchief combined with flexion of the thigh on the belly, and of the leg on the thigh.

For **Hemorrhage from the Arteries of the Leg,** pressure should be applied over the course of the large blood-vessel, which is superficial in the hollow behind the knee-joint. This pressure may be made by means of the elastic bandage, or by the knotted handkerchief combined with flexion of the leg on the thigh and of the thigh on the abdomen.

For **Hemorrhage from Arteries of the Foot,** the bandage and pad may be applied, or the above measure may be employed.

Internal Hemorrhage, by which is understood any bleeding in the great cavities of the trunk in which the blood accumulates internally, may be suspected when, after injury, liable to cause it, symptoms of faintness from loss of blood come on (*vide* Fainting). *It is to be treated by rest in the* *recumbent position, the head lying lowest, and cold applications externally.*

Fainting is one of the results of copious hemorrhage. If it occurs as the result of excessive hemorrhage, the bleeding should first of all be controlled. *The patient should be laid on his back, with his head low; his arms and feet may be elevated.* Tight clothing should be loosened and cold water sprinkled upon the face. The application of warmth to the body, the holding of hartshorn to the nostrils, and the careful administration of stimulants (when the patient is able to swallow) may be necessary.

It must be borne in mind that with the advent of a fainting spell and its attendant weakening of the heart's action a cessation of hemorrhage often takes place naturally; hence measures are to be adopted to prevent a recurrence of the bleeding when reaction comes on.

In fainting from internal hemorrhage the treatment outlined above will in the main be applicable, *stimulants, if given at all, being administered with extreme caution.*

It may here be mentioned that fainting is due to numerous causes besides hemorrhage. Debility in general, fright, and nervous impression of various kinds may produce it.

FIRST AID TREATMENT OF WOUNDS.

A **Wound** is any breach in the soft tissues of the body caused by violence. Upon the battle-field the following forms of this class of injury will present themselves, viz., the *contused*, the *incised*, the *punctured*, the *lacerated*, and the *gunshot wound*. *Contusions*, though not properly wounds, may advantageously be considered with them.

Contusions, or *bruises*, are caused by blunt, heavy instruments. On the battle-field they result from falls, blows with the butt of a gun, the passage of a wagon or piece over a portion of the body, etc. Ordinarily, they are not of serious nature; but complicated with injuries of internal organs or fractures, their import is much graver. Simple contusions are treated with applications of cold water or of laudanum. Hot, wet applications are of use where the pain has ceased; they favor the absorption of the blood which has escaped into the tissues. If the bruise is complicated by fracture, the latter should be attended to in a manner to be described hereafter. If the patient is unconscious from a blow on the head, the bearers should not try to bring him to his senses by shaking him, but should sprinkle cold water upon his face and chest for that purpose, and place him under care of the surgeon without delay.

A **Contused Wound** is a bruised one, and as far as its management in general is concerned, it may be considered as a species of contusion. Ordinarily, a moist, cold dressing of lint with a compress and bandage suffices. Should hemorrhage exist, not controllable by the pressure of the dressing, the elastic bandage, windlass, tour-

FIRST AID TREATMENT OF WOUNDS. 35

niquet, etc., should be employed. The complications of the contused wound are those of the contusion.

When the word *lint* is employed herein in connection with the dressing of wounds, either the picked, shredded, or sheeted article, or its cotton substitutes, are to be understood. Preferably, all dressings should be antiseptic. The antiseptic compress of the Esmarch's package serves the purpose both of lint and ordinary compress.

An **Incised Wound** is a cut such, for instance, as is made by the sabre. Hemorrhage may be slight, in which case a dressing of lint and a compress, covered by a roller or triangular bandage, snugly applied, will check it. Should it be severe, recourse, in addition, must be had to the elastic bandage or other appliance of its class; or pressure with the fingers, as in wounds of the carotid artery, may have to be employed. *Patients with vertical cuts of the walls of the belly should be carried lying on their backs, those with horizontal cuts are to be transported seated.* Incised wounds of the belly are occasionally followed by protrusion of the bowels. If not injured or soiled, gentle efforts should be made to return the intestines. If this is unsuccessful, they should be covered and protected by a clean cloth while the patient is conveyed to the dressing station.

A **Punctured Wound** is made by a stabbing weapon, such as a bayonet. Generally there is but little hemorrhage, and the ordinary dressing is sufficient for the purposes of exerting pressure.

Fragments of the weapon left in the wound should not be removed by the bearer, but by the surgeon. If the lungs have been injured, which in some cases may be suspected from the location of the wound, and the coughing up of blood, etc., the patient is neither to be conveyed in a wagon nor allowed to walk to the dressing station, but must be carried thither on a stretcher.

A **Lacerated Wound** is one in which the parts have been torn. The ragged stump left when a portion of the hand has been wrenched off by a premature explosion is an example. The wound should be cleansed and foreign bodies removed by a gentle stream of water from a sponge; the parts should be restored to their natural positions as far as possible, and a cool, wet cloth, or one moistened with laudanum or alcohol, is then to be applied and covered with the bandage. If hemorrhage is severe, the tourniquet or other means for arresting it may have to be employed. If the injury is great and the patient is suffering from shock, stimulants may be called for. Extensive lacerations should *not* be treated with cold applications; where hot ones are not available, dry dressing or cloths moistened with alcohol or laudanum are preferable.

Gunshot Wounds are produced by missiles projected from weapons loaded with explosive material. These missiles include bullets, gunshot, shrapnel, shells, grape and canister shot, etc. Gunshot wounds involving merely the skin and muscles are trivial, but such as are complicated with injury to the internal organs or with fractures, are among the gravest of the casualties of war. The amount of bleeding attending these cases is variable. The management of gunshot wounds depends upon whether they are simple flesh wounds, or are complicated with hemorrhage or fracture. Flesh wounds, after thorough cleansing, are dressed with lint, compress, and bandage. Hemorrhage, if serious, requires the tourniquet, or other means of this class, to control it, while fractures are to be treated in a manner to be described hereafter. Frequently, when a portion of a limb has been torn off by a fragment of a shell, there is little or no bleeding from the arteries, because the wrenching force has twisted the arterial coats in such a manner as to close the vessels. They are not so firmly closed, however, but what a jar may start the hemorrhage; hence, as a safeguard, the tourniquet is loosely adjusted over the course of the main artery of the limb. Gunshot wounds generally have an opening of entry and one of exit; in case the projectile fails to emerge, the latter, of course, is absent. Bearers should not try to remove imbedded projectiles.

The *first aid package* of Esmarch will be found to be available as a first dressing for all classes of wounds.

Shock is a form of collapse accompanying severe injuries. Fright, despondency, hunger, thirst, and debility favor its development, while loss of blood aggravates it. The symptoms are those of utter prostration. The skin is pale, cold, and clammy; the breathing shallow, the pulse feeble; the eyes are dull, the eyelids drooping, and the pupils dilated. The patient's mind wanders, and unconsciousness sometimes comes on. Well-marked shock is a serious condition, and demands prompt measures. The patient is to be laid down, head low, and covered with blankets, all hemorrhage having been checked. Hot applications (hot bottles, hot plates, etc.) should be made to the whole body, and especially to the region of the heart and the pit of the stomach, care being taken that the heated articles are not laid directly upon the bare skin. Hot drinks should be given, or a teaspoonful of brandy in a tablespoonful of water administered every ten minutes for several hours. When, however, there is suspicion of internal hemorrhage, liquor had better be omitted, unless given with extreme caution.

FIRST AID TREATMENT OF FRACTURES.

Fracture.—When a bone is broken a *fracture* is said to have taken place. Fractures may be classed as *simple* and *compound:* in the former the skin is not broken, in the latter it is. A compound fracture is a very much graver injury than a simple one. That a bone has been broken can be recognized by the occurrence of pain, by the faulty position of the limb, by its bending where it ought not to, and by the so-called *crepitus*, or grating sound or sensation, produced by motion at the seat of injury. Pain is usually not marked when the limb is quiescent, but the patient's efforts to move it, or unskilful handling thereof, cause acute suffering. It is the duty of the bearers to support the affected limbs by means of splints, and to keep the fracture immovable. A rectification of the faulty position of the fragments of the bones is not to be attempted by them, unless excessive pain calls for such a measure. To effect it, one man grasps the limb above the fracture and the other below, both pulling in opposite directions until the proper shape is restored. Care must be exercised not to increase the damage already done.

Extemporary Splints, made from a variety of material available on the battle-field, are utilized by the bearer in *putting-up* fractured limbs for the time being. The articles from which they are improvised comprise rifles, side-arms, coats, capes, knapsacks, straw-mats, rolls of straw, boards, pickets, sticks, bark, laths, switches, rushes, leather straps, telegraph wire, strips of tin, and many others. Temporary splints may be applied directly to the limb or over the clothing, according to circumstances. Their mode of application will be shown in the consideration of the following fractures:

Fractures of the Upper Arm may be put up with a bayonet and scabbard, one on each side, or with bundles of straw lined with soft material. Again, thin pieces of board may be applied to the inside and outside of the limb, and, if necessary, to the front and rear also. In all cases the splints are firmly held together with triangular or other bandages, and the fore-arm is flexed and carried in a sling. The simplest method is to utilize the chest as a splint. The elbow is drawn down to the side of the body, a layer of cotton or linen cloth being interposed. The whole upper arm is then to be bound to the chest, the fore-arm being carried in a sling, the hand a little higher than the elbow.

Fracture at the Lower End of the Arm.—This may be put up with a rectangular inside splint in order that more support be afforded than can be given by the straight splint. The rectangular splint may be improvised from telegraph-wire, and should be well padded.

The outside of one of the blades of the ordinary straight scissors should be provided with a file edge, by which a notch may be cut in the wire to weaken it and cause it to break readily.

Fracture of the Fore-arm is put up with two light splints; shingles are well adapted for the purpose. The fore-arm being flexed and the thumb turned upwards, the splints, well padded and long enough to extend beyond the fingers, are applied to the front and rear of the limb, and firmly bound on; the hand is carried in a sling and raised slightly higher than the elbow.

Whenever a sling is required for the fore-arm, one may be conveniently made by turning up the bottom of the blouse, passing it over the limb, and pinning it to the breast of the garment. If necessary, the side seam may be slit up.

Fracture of the Finger may be put up with a light splint reaching from wrist to finger-tip, the finger being straightened out.

Fracture of the Thigh may be put up with a splint placed on the outer side of the limb and extending from axilla to foot. For this purpose a rifle may be used, with the butt placed in the armpit and the hammer pointing towards the ground. The weapon is secured by a bandage passing around the foot and another passing around the trunk, each being made to encircle the splint. A soldier's overcoat is then adjusted so as to constitute a pad on one side of the limb, and passing under it, to form a stiff roll on the other, the whole being fastened by straps, bandages, handkerchiefs, etc.

Fracture of the Knee-Pan may be put up with a straight splint extending from hip to heel on the outer side of the limb. A picket may be used, if available.

In fractures of the lower extremity, in general, the limb should not be held absolutely straight, but slight bending at the knee should be allowed.

In **Fracture of the Leg** the limb may be carefully drawn down and placed in a natural position. An overcoat or cape should then be passed under it and made to form a stiff roll on each side. A piece of board, a pair of bayonets, or a sword, etc., is then laid along the limb over the coat and secured by bandages, straps, etc., or the other leg may be utilized as a splint to which the injured one is bound.

Fractures of the Spine is difficult to detect. When suspected, the patient should be laid upon his back, and left for the surgeon's examination. He is not to be disturbed more than is absolutely necessary.

Fractures of the Shoulder-Blade, Hip-Bone, or Ribs are frequently hard to make out; the first two are rare. First-aid treatment consists in placing the sufferer in a comfortable position and securing rest and coolness.

In **Fracture of the Collar-Bone** the patient should be laid flat on his back without pillow, the arm bound to the side and the fore-arm secured in a sling made of the turned-up bottom of the blouse pinned to the breast of the garment.

Fracture of the Skull calls for the recumbent posture and cold applications to the head. Further than attending to these requirements and to the comfort of the patient, the bearer will do nothing.

In **Fracture of the Jaw** the mouth is to be closed and a bandage applied to keep the teeth together.

Compound Fractures, after the wound has been dressed, are to be put up in like manner as the simple forms, care being taken that the fragments cause no further damage to the soft tissues.

FIRST AID TREATMENT OF DISLOCATIONS.

Dislocations consist in the displacement of the articular extremities (joint ends) of bones from their sockets. The bearer, as a rule, should not attempt their reduction, but should place the limb in a comfortable position, and after applying cold dressings, secure it by bandages.

In the service, dislocations are often caused by men being thrown from horses, caissons, etc.

FIRST AID TREATMENT OF SPRAINS.

A **Sprain** is caused by the overstraining of the muscles and ligaments covering a joint, the force exerted stopping short of that required to produce a dislocation. The wrist and ankle are joints most frequently affected by this class of injury. Severe sprains sometimes present themselves on the field of battle. Their treatment resembles that of contusions; cold dressings and, if necessary, splints are applied; later, hot applications should be employed to produce absorption.

PROCEDURES TO BE ADOPTED IN CASES OF SUSPENDED ANIMATION.

Often, as a result of the shock of severe injuries, profuse bleeding, or various forms of obstruction to breathing, a trance-like condition known as *suspended animation* arises, characterized by cessation of respiration, and an all but complete failure of the heart's action. In order to avert death in a case of this kind, the bearer, in addition to the measures applicable in fainting and shock, will employ what is termed *artificial respiration*.

By **Artificial Respiration** is meant a procedure in which, by alternate compression and expansion of the chest, the lungs are made to imitate the natural action of breathing. There are several methods of artificial respiration, the best known among which are Silvester's, Marshall Hall's, and Howard's, the first being the simplest, and readily carried out by one man. It will be more advantageous for the bearer to confine himself to one of them, learning it thoroughly, than to endeavor to master them all. Silvester's method is hence selected for the purpose of description. It may be occasionally necessary to employ artificial respiration for hours; it is discontinued when natural breathing recommences, or the case is finally given up as hopeless.

In **Silvester's method** the patient is laid upon his back, with the arms stretched out along the side of the body, and a firm roll of some kind (clothing, blanket, etc.) placed under the shoulder-blades for the purpose of raising the shoulders and extending the throat. The tongue, which has a natural tendency to fall back and obstruct the wind-pipe, is drawn forward and secured, either by a string, etc., passed around its base and the chin, or in various other ways that may suggest themselves. Kneeling behind the patient's head, the bearer seizes the arms above the elbows, and draws them outwards and upwards until they are fully extended above the head. After a pause of about two seconds the arms are carried back to their original positions, the bearer making firm pressure upon the chest at the same time. This procedure is carried out at the rate of about fifteen times a minute. Whenever the arms are raised the chest is expanded and air enters the lungs; when they are brought down and pressure is made upon the chest the latter is compressed and air is expelled. The natural movements of respiration are hence imitated.

PART III.

Management by the Bearer of Ordinary Accidents and Emergencies.

In the preceding pages the scope of the duties of the bearer on the field of battle have been outlined; but besides ministering to the needs of those wounded in combat he should furthermore be able to afford first aid in the accidents and emergencies of ordinary garrison life, and to take care of minor cases when the surgeon's services are not available.

GENERAL RULES TO BE OBSERVED IN CASES OF ACCIDENT.

Should the bearer happen to be present at the scene of an accident at which his services are called upon, he should first of all request bystanders, who may have crowded around, to press back in order that breathing-space and room for manipulation be obtained. If he needs assistance, he will ask for as few helpers as he can get along with. The patient is placed in a comfortable position upon his back, tight clothing loosened, and the nature and extent of his injuries are determined. The examination completed, word is sent to the surgeon and the hospital steward, the character of the injury being stated for their guidance in the matter of instruments and appliances necessary. In the mean time the bearer will adopt such first-aid measures as the situation may call for, being careful to observe the limitations of his functions.

HEMORRHAGE, WOUNDS, FRACTURES, DISLOCATIONS, SPRAINS.

These accidents having already received consideration in Part II., need not be discussed here.

CONDITIONS CAUSING LOSS OF CONSCIOUSNESS.

Loss of Consciousness occurs under a variety of circumstances, the conditions causing it being at times readily discernible, at others difficult to discover. It may be due to fainting, shock, concussion, compression, apoplexy, or other disorder of the brain, sunstroke, heat-exhaustion, intoxication, epilepsy, poison, or a number of other troubles. It is not to be presumed that the bearer will be able to make what the surgeon calls a *differential diagnosis* between the various conditions producing unconsciousness, that is to say, to ascertain absolutely whether insensibility results from this, that, or the other cause. Professional study is necessary for such purpose; it is not within the scope of the bearer's duties. He will mainly have to rely on what he may learn from the patient's associates, and on circumstances which the intelligent layman may reasonably be expected to observe. When the bearer is in doubt as to the origin of unconsciousness, as he may well be in many cases, he must rely upon general principles. In all instances he must try to speedily obtain the surgeon's services. Pending the latter's arrival, he should lay the patient upon his back, with his head slightly elevated, and loosen tight clothing. If there be pallor, clamminess of surface, shallow breathing, and weakness of pulse, heat should be applied to the body, hartshorn held to the nose, and hot drinks given as soon as the patient is able to swallow. If the skin feels excessively hot, cold applications should be made to the head and body, and cold drinks given when consciousness returns.

Fainting and Shock have already received attention in a preceding chapter.

Concussion of the Brain or stunning is caused mainly by blows or falls upon the head. It is accompanied by feeble pulse and vomiting, pallor, depression, incoherence of language, and partial or complete insensibility. The proper plan of treatment is to lay the patient upon his back, loosen tight clothing, and secure quiet and fresh air. Heat should be applied if the skin becomes cold and clammy. Stimulants had better be avoided.

Compression of the Brain is due either to pressure exerted upon the organ by a fragment of a broken skull-bone, or by blood which has escaped into the cranial cavity from a vessel ruptured as a result of external violence. The symptoms are insensibility, slow pulse, snoring, breathing, paralysis, muscular twitching or convulsions, and dilatation of one or both pupils. The recumbent position should be secured for the patient, and cold applications made to his head; beyond this the bearer will attempt nothing without medical advice.

Apoplexy is due to the rupture of a diseased blood-vessel in the brain without external violence. Usually the patient's face is flushed, and may be drawn to one side. The loss of consciousness may be gradual, or insensibility may come on suddenly. The pulse is slow, the breathing snoring. The bearer will adopt the same measures as in *compression of the brain*.

Sunstroke is due to overheating of the body, associated with an inability of the skin to perform its proper functions on account of the influences of a close atmosphere. Its symptoms of warning are headache and oppression. These after a time are followed by loss of consciousness. The breathing is labored, *the skin intensely hot, perspiration absent;* the bladder and bowels sometimes discharge involuntarily. The great aim in treating this most serious condition is *to reduce the bodily temperature as soon as possible.* The patient should be immediately conveyed to a cool, airy place, and on removal of his clothing placed in a cold bath, or wrapped in a wet-pack, that is to say, in sheets soaked in water, and which in this case are kept wet by repeated applications of cold or iced water. If these methods are impracticable, cold must be employed in some other way, as by thorough and continued sponging of the body and head, lumps of ice being rubbed over the chest and placed over the large blood-vessels in the armpits and groins. If consciousness returns, the application of cold should be discontinued, to be renewed only if the temperature again rises above normal, or insensibility comes on once more.

Heat Exhaustion is a prostration of the system due to excessive heat, but is not accompanied by loss of the transpiratory function of the skin.

It resembles the ordinary *fainting spell*, and is to be similarly treated, the patient being removed to a cool, airy spot. Unlike sunstroke, this condition presents a moist, cool skin.

Intoxication in its fully developed symptoms somewhat resembles *apoplexy*. When the bearer is certain that the case is one of drunkenness, an emetic composed of a teaspoonful of ground mustard stirred up in a teacupful of lukewarm water may be given. This is to be followed by a teaspoonful of aromatic spirits of ammonia in a similar quantity of water, or a large draught of vinegar, after vomiting has occurred. If the patient be, as is called, dead-drunk, and in danger of dying from the collapsing effects of the liquor, the general application of heat to the body is imperative. Emetics should not be employed if the bearer is not absolutely certain of the nature of the case, as in apoplexy they would create irreparable damage. In cases of doubt the precautions necessary in *apoplexy* should be adopted.

Epileptic Seizures, in the main, should be treated like fainting spells. No attempt should be made to violently prevent the spasmodic movements: they should simply be controlled. To forestall biting of the tongue, a folded towel or some other available article should be thrust between the teeth,—not in such a manner, however, as to interfere with breathing. After the seizure is over, rest in the recumbent posture is necessary.

Poisons, such as opium and chloral, produce insensibility when taken in overdose. If the bearer knows that these drugs have been used, he will adopt the first-aid measure to be hereafter detailed as applicable in cases of poisoning; often, however, the cause of the patient's unconsciousness will be unknown to him. In all cases, however, he will lose no time in communicating with the surgeon.

ASPHYXIA.

Asphyxia is a term applied to all accidental conditions in which life is in danger on account of any obstruction whatsoever to respiration. The conditions mentioned include suffocation from drowning, strangulation, noxious gases, foreign bodies in windpipe or gullet, etc.

In **Drowning,** *if the patient hass topped breathing,* tight clothing is first of all loosened, the individual is then turned over on his face, a roll of clothing, a rolled blanket, etc., being placed under his stomach, his mouth and nose are cleared of sand, mud, or other substances collected therein, and pressure is made upon the spine and kept up until water ceases to flow from the mouth. The patient is then turned over on his back, and the roll placed under his shoulder-blades so as to raise the shoulders and extend the throat. The tongue being drawn well forward, is either secured by a string or rubber band passing around the base of the organ and the chin, fixed by thrusting a small stick or pencil across the top of it behind the molar teeth, or held by an assistant. These procedures accomplished, Silvester's method of artificial respiration should be practised in the manner indicated on page 44. It should be kept up for hours if necessary, and until natural breathing returns, or the case has been given up as hopeless by competent authority.

While the bearer has been busied with the drowned man, bystanders have, by his direction, been sent to obtain warm and dry coverings to replace the wet clothes of the patient. A fire should be built by others, by which water, bricks, stones or pieces of metal may be heated, and blankets, coverings, clothes, thoroughly warmed. An assistant, without interfering with the process of artificial respiration, pulls off the wet clothes, replaces them by warm coverings, applies heat to the body in the shape of hot bottles, hot bricks, or hot stones, properly covered, or even the hot sand of the beach. The body and limbs should be constantly rubbed towards the heart.

As soon as the patient is able to swallow, a teaspoonful of hot liquor in a tablespoonful of water may be given every few minutes until the danger is over.

As soon as the patient begins to breathe of his own accord, the artificial process should be timed to aid the natural respiration. Breathing may be stimulated by holding hartshorn to the nose, slapping the skin, or by dashing hot water upon the chest.

As a rule, the drowned man should never be removed during the employment of methods for his resuscitation. After he has been, so to speak, brought back to life, he should be cautiously carried away in the recumbent position, put in a warm bed, and carefully watched for stoppage of breathing.

If the patient has not stopped breathing when drawn out of the water, the procedure is similar to that employed in the preceding case, artificial respiration, however, being omitted except when the natural function begins to fail.

Strangulation from Hanging, etc., is treated by removal of the obstruction, stimulating natural breathing, and, if necessary, by performing artificial respiration.

In **Suffocation with Gases** the patient is first of all to be carried into a pure atmosphere. Breathing is to be stimulated by sprinkling the face with cold water, tickling the throat and nostrils with a feather, the application to the nose of ammonia, etc. If necessary, artificial respiration may be employed.

Caution is necessary in the rescue of these patients from their dangerous surroundings, and the bearer should take adequate measures to protect himself. In entering a room filled with illuminating gas he should be careful to carry no light. When the patient is to be hauled up from a well or vault, the gas in it may, in a measure, be dissipated by throwing down a few buckets of water. A cloth, veil, or like covering worn over the head will prove of some value to the rescuer in keeping the gas out of his lungs.

Suffocation from Foreign Bodies in the Windpipe or Gullet is to be treated by a prompt removal of the obstruction, and, if necessary, the subsequent employment of artificial respiration. Substances clogging the windpipe cause a good deal of coughing : those obstructing the gullet interfere with swallowing, but do not give rise to cough. If the foreign body can be reached, it should be pulled out by the fingers, a bent hairpin, a pair of blunt-pointed scissors, etc. Sometimes blowing forcibly into the ear causes a reflex cough, which is sufficient to dislodge the body. Foreign bodies in the windpipe are generally coughed up ; if not, the case will require the surgeon's services. Coughing may be aided by blows on the back, and a quick compression of the chest by the hands. Inversion (turning upside down) of the body may dislodge the substance. When breathing is not interfered with and the foreign body is not accessible, the bearer will limit himself to causing the patient to lie down, and making him as comfortable as possible pending the surgeon's arrival.

In connection with the above, and as coming within the scope of the bearer's duties to carry out a line of simple treatment in absence of the surgeon, it may here be mentioned that after a foreign body has been swallowed, and particularly if it be one with sharp points, a generous solid diet, including a liberal allowance of vegetables, should be given, in order that the substance may thus be imbedded in the solid waste of the intestinal canal. No purgative should be administered.

BURNS AND SCALDS.

In **Burns and Scalds** the first thing to do is of course to remove the cause. Should clothing have caught fire, the individual is to be thrown down on the ground and deluged with water or rolled in an available rug, blanket, or coat. Great damage arises from the inhalation of flames, hence in wrapping up a patient, the bearer should first cover the upper part of the body. When clothing is removed before dressing the injury, adherent portions should be carefully cut out of the garment and left. The dressings required and the further treatment necessary are given in the following paragraphs.

Slight Burns or Scalds are best treated with cold applications. Involving an extremity, the limb may be entirely immersed in water. A solution of bicarbonate of soda (baking-soda), composed of a heaped tablespoonful of this substance to a teacupful of water, soon relieves pain; the soda may also be used in bulk, slightly moistened. White of egg, carron-oil (lime-water and linseed-oil mixed in equal parts), and lather applied with a shaving-brush are also useful.

In **Burns and Scalds causing Blisters** the latter are to be carefully opened with needle or scissors. Applications of cold water, fresh leaves, thinly sliced or grated potatoes, white of egg, or carron-oil will be found of value, the great object of these dressings being to exclude air. Substances which crust or cake should be avoided.

Deep Burns or Scalds may first be treated with a water dressing, preferably. The surgeon should be called upon for a further management of the case.

Shock occurring as a Result of an extensive Burn or Scald should be treated in a similar manner to shock in general.

FREEZING.

In cases of **Freezing,** a warning symptom of danger is an uncontrollable desire to lie down and sleep, which the sufferer's companions should do their best to prevent. On reaching a place of shelter and treatment, the sufferer must be gently handled, as otherwise parts of the body such as fingers, toes, the ears, are liable to be broken off. He must never be brought immediately into a warm apartment, but efforts at bringing him to must be made in a cold room. The clothes must be cut off, if necessary. The patient must be rubbed all over with cold water, snow, or put in cold wet sheets, until he becomes limber enough to justify removal to a cold bed. Artificial respiration is now to be employed, if called for, and warm drinks (not stimulants) are given as soon as he is able to swallow.

Frost-Bites, that is to say, local freezing of exposed portions of the body, are to be similarly treated as to friction with cold water and snow. The same cautions obtain as to sudden exposure to warmth. Very often a person afflicted with frost-bite is unaware of his trouble, as the affected part has utterly lost all feeling. A passer-by will recognize the dangerous condition of the organ, supposing it is the ear or nose, by the blanched, waxy appearance it presents. Prompt measures should be taken immediately to prevent total loss of the member.

Chilblains are due to a chilling of the circulation of certain parts, particularly the toes. They are attended with a good deal of discomfort in the way of itching and smarting, especially manifested after the patient has gone to bed. A good plan to avert these symptoms is to generally keep the affected parts away from the fire, and to bath them in cold water or rub them with snow before retiring. Stimulating ointment may then be applied, as, for instance, the compound ointment of resin.

SORENESS OF THE FEET.

Soreness of the Feet, in those unaccustomed to marching, may be avoided by soaping or greasing the feet thoroughly before setting out. The march made, they should be washed or wiped clean and dry. The feet may be rendered tough by soaking them in strong tepid solutions of alum or tannin.

The German Fussstreupulver is an excellent

preventive of sore feet. Sifted in shoes and stockings, it keeps the feet dry, prevents chafing, and heals sore spots. It is composed mainly of soapstone, to which starch and salicylic acid have been added (87 parts soapstone, 10 starch, 3 salicylic acid, by weight).

Blisters should be punctured at the lowest point, and the fluid allowed to drain.

BLEEDING FROM THE NOSE, LUNGS, STOMACH, OR BOWELS.

Nose-Bleed is often salutary, being a natural method of relief in rush of blood to the head. Occasions may arise, however, when it is so copious as to create alarm and necessitate the surgeon's services. In this emergency the bearer should, pending the latter's arrival, endeavor to check the flow by causing the patient to snuff up vinegar, or, in default of this, a solution of alum and salt, or by plugging the nostrils with styptic or other form of cotton. Pressure should likewise be exerted upon both *facial arteries:* this may be accomplished by pushing the fingers firmly against the lower jaw-bone, on each side, immediately in front of the lower part of the muscles, which are prominently visible at the angles of the jaws when the latter are firmly clinched.

Hemorrhage from the Lungs proper is rarely copious. The blood is *coughed* up, and is usually bright red and frothy from contained air-bubbles. When this form of bleeding occurs, the patient should be put to bed, and bolstered up in a sitting position. Cold drinks should be given, and if the patient is not too weak, cold applications are to be made to the chest. Salt and vinegar are popularly supposed to be efficacious in checking pulmonary hemorrhage, and may be given in the following doses every fifteen minutes: vinegar, one teaspoonful; salt, about a quarter of a teaspoonful.

In **Hemorrhage from the Stomach** the blood is *vomited* up, and presents a characteristic appearance of coffee-grounds, from having been acted upon by the digestive fluids. Mixed with food, it may be difficult to recognize. The first-aid treatment in these cases calls for rest in bed, cold drinks, or bits of ice, vinegar in teaspoonful doses, and cold applications to the belly.

In **Hemorrhage from the Bowels** the patient should be put to bed, injections of ice-water given, and cold applications applied to the belly.

POISONING.

By the term *poison* is meant any substance whatsoever which, taken into the system in small quantities, is capable of producing noxious or deadly effects.

Poisons may be classed as *corrosive*, *irritant*, and *neurotic*. Corrosive poisons burn the parts with which they come in contact; irritant poisons cause intense inflammation of such parts; neurotic poisons manifest their action chiefly through the nervous system.

An *antidote* is a remedy counteracting the effects of a poison, either by uniting with it and forming a harmless compound, or by exerting an action upon the system which is opposed to and neutralizes that of the poison. The former, mainly, will concern the layman. The general principles to be remembered in all cases of poisoning are, that the poison is to be neutralized and removed as soon as possible; that any irritant or caustic action it may have is to be counteracted by demulcent (bland, soothing) fluids; that if collapse or unconsciousness threaten, proper measures are to be adopted to combat the tendency; and that artificial respiration is to be employed if necessary.

It is impracticable within the limits of this manual to detail the first-aid treatment of each and every case of poisoning that may occur. The above rules, however, should be firmly fixed in the mind of the bearer. Should he be unacquainted with the antidote to be applied in any given case, he should not lose confidence, but do his best in carrying out the other measures indicated.

To illustrate the general measures mentioned, the treatment in the case of certain common poisons will be detailed in full. Such procedures only will be given, however, as the layman may be expected safely to employ. The administration of such antidotes as are themselves powerful poisons will not be advised; it should be left to the surgeon.

It should not be forgotten that when poisoning has occurred the surgeon's services should be procured as soon as possible.

General Measures and Remedies.—Under the head of the various poisons, certain remedies and measures of general application will be constantly referred to; in order to avoid the necessity of detailing them in full each time, it will be well to consider them before taking up the subject of the individual poisons. The topics in question include the use of emetics, certain antidotes, bland liquids, stimulants, and laudanum.

Emetics.—Vomiting may be produced by tickling the throat with a feather or the finger, or by administering warm water, solution of salt, mustard, sulphate of zinc, ipecac and water.

Tickling the throat with a feather or the finger will frequently bring on vomiting. When an emetic has already been given, this procedure will hasten its action.

Warm water is to be given in as large quantities as a pint at a time, repeated at intervals of a minute or so.

Salt may be given in as strong solution as can be made with water, a teacupful of the brine being administered every minute or two.

Mustard is given with water in the proportion of one tablespoonful to the pint of fluid. Copious draughts of tepid water assist the emetic action.

The mustard, moreover, exercises a stimulating influence on the system.

Sulphate of Zinc (white vitriol) is a valuable emetic. It is given in twenty-grain doses (as much as may be heaped on a silver quarter), dissolved in water. This dose should be followed by a cup of tepid water, and repeated every three minutes until four doses have been given, or vomiting has occurred.

Ipecac is given with water in the proportion of a teaspoonful of the powder or a tablespoonful of the syrup to a pint of fluid.

Alkaline Antidotes are employed to form compounds, harmless, or comparatively so, with acids in cases of acid-poisoning. They include ammonia diluted (a tablespoonful to two teacupfuls of water), lime-water, solution of baking-soda, bicarbonate of potash, or soap and either or several of the following articles mixed with water, viz: chalk, whiting, whitewash, tooth-powder, plaster from the walls, wood ashes, etc. In preparing the mixture, no time should be lost in endeavors to make a perfectly uniform one. A little grit will do no harm.

Acid Antidotes are employed to form harmless or comparatively harmless compounds with acids in cases of poisoning with the latter. Those which the bearer should especially remember are vinegar and lemon-juice.

Bland Liquids are employed to soothe the membranes which have been irritated or corroded by the action of a poison. They include beaten eggs, oil, milk, condensed milk, gruel, barley-water, arrow-root, flour and water, starch decoction, mucilage, etc.

Stimulants may be necessary in cases of threatening collapse or unconsciousness. They include liquors, wine, tea, coffee, and dilute ammonia (teaspoonful to a teacupful of water). Tea and coffee are furthermore valuable as being antidotes to several poisons, on account of the tannin they contain. In preparing them no time should be lost in settling or straining. Whiskey may be given in teaspoonful doses, with tea or coffee, or as a toddy. Stimulants may be given by the bowel.

Laudanum may be given in a dose of twenty-five drops in case of severe pain. This dose may be repeated in thirty minutes if the pain persists and drowsiness has not been caused.

Forcible Administration of Remedies.—It may sometimes happen that the patient through perverseness or fright will refuse to open his mouth and swallow the remedies tendered, or he may be prevented by unconsciousness from so doing. In either case he should be laid upon his back, and the bearer should insert both thumbs into his (the patient's) mouth, between the cheeks and the teeth and along the line of the edges of the latter, slipping the tips of the thumbs into the space behind the last tooth on either side. The jaws can then be readily separated, the bearer running no risk of having his fingers bitten. The handle of a strong spoon, a stick, a paper-cutter, or other available article should then by a second bearer be thrust well back upon the tongue, and used as a tongue-depressor. The fluids required may thus be poured down the patient's throat without trouble, provided the tongue be sufficiently depressed.

In **Poisoning from Unknown Substances** the stomach should be cleared by vomiting, incited again and again. Subsequently bland and soothing drinks may be given. Pain may be relieved by laudanum, in doses as prescribed on page 63.

Collapse, if it occur, is to be treated like collapse in general. The patient is to be put to bed, warmth applied in the form of hot bottles, bricks, or stupes, and stimulants and hot drinks (tea, coffee) are to be administered. A teaspoonful of whiskey may be given every ten minutes, either in hot water, or in hot tea or coffee.

Corrosive Poisons include the mineral acids—sulphuric (oil of vitriol), nitric, and hydrochloric (muriatic); oxalic acid, carbolic acid ; the alkalies—ammonia, soda, and potash ; corrosive sublimate, nitrate of silver (lunar caustic), phosphorus, and various salts of different metals.

The action of the mineral acids and the alkalies is especially rapid and violent, and when they have been taken undiluted, little is to be confidently expected to result from the treatment adopted.

In all cases of poisoning by corrosives, it is to be understood, without further specifying, that rest is to be secured, and stimulants administered, if necessary ; that collapse is to be treated as indicated in the preceding paragraph; and that pain is to be relieved in the manner detailed on page 63.

Sulphuric, nitric, and hydrochloric acids are neutralized by alkalies. The alkaline antidotes mentioned on page 62 are all useful in cases of poisoning with these substances, and should be freely administered.

After neutralization, vomiting should be produced, and bland and soothing drinks subsequently given.

Carbolic acid, in cases of poisoning by it, calls for repeated emetic draughts and large quantities of demulcent drinks.

Oxalic acid is to be neutralized by lime, as with soda, ammonia, or potash salts it forms compounds which themselves are poisonous. Chalk, lime, whitening or whitewash, stirred up in water, or lime-water itself, may be plentifully administered. Magnesia also may be used. After thorough neutralization, vomiting should be brought on, and soothing drinks given.

Ammonia, soda, and potash, the *Alkalies*, are to be neutralized by dilute acids. Vinegar is to be given in as large quantities as a pint at a time, if possible. Lemon-juice may be of value in minor cases. After thorough neutralization, vomiting should be excited, and demulcent drinks, especially olive-oil, subsequently administered.

Lye, a preparation containing alkalies, and *ammonia liniment* are sometimes a source of alkali-poisoning.

Corrosive sublimate is both neutralized by

white of egg (raw) in large quantities, or tannin, conveniently given in the form of strong tea. Vomiting should be produced as soon as possible, and eggs and milk subsequently administered.

Nitrate of silver, *lunar caustic*, is to be neutralized by common salt, given in solution and in quantities large enough to produce vomiting at the same time. Mustard or ipecac may be also used as emetics. Subsequently bland drinks should be administered.

Phosphorus.—Poisoning by phosphorus is to be treated by the administration of calcined magnesia and water, followed first by an emetic and later by demulcent drinks. *No oils or fats are to be given*, as they facilitate the absorption of the poison.

Irritant Poisons may be classed as *metallic*, *vegetable*, and *animal*. Examples of metallic irritants are arsenic, certain antimony, lead, and zinc salts, iodine; of vegetable, croton-oil, the essential oils, gamboge, etc.; of animal, cantharides (Spanish fly).

In all cases of irritant poisoning it is to be understood that rest is to be secured, and stimulants, when necessary, are to be administered, that collapse is to be treated and pain relieved, according to the principles already laid down.

Arsenic.—In poisoning by arsenic or any of its preparations (*Paris green*, etc.) vomiting should first of all be induced. Demulcent drinks are to be freely administered, as they dilute the poison, protect the stomach, and facilitate vomiting. Dialyzed iron in repeated ounce doses (two tablespoonfuls) is to be given as an antidote, each dose being followed by a teacupful of brine. In default of dialyzed iron, equal parts of the sulphate of iron (green vitriol) and carbonate of soda may be dissolved in separate cups of hot water, mixed and administered. Large quantities of calcined magnesia in water may be used if the iron preparations are not obtainable. After thorough neutralization of the poison, final vomiting should be induced, a dose of castor-oil given, and soothing drinks administered.

Tartar emetic is to be neutralized by tannin administered in the form of the powder itself, strong tea, nut-galls, or powdered oak-bark. One of the prominent symptoms of poisoning by tartar emetic is vomiting; this should be assisted by copious draughts of tepid water or demulcent drinks. A teaspoonful of tannin in water, strong green tea, or half a dozen nut-galls, powdered in water, etc., may then be given as an antidote. After the poison has been *neutralized* and the stomach thoroughly cleared out, demulcent drinks should be administered.

Lead.—Lead-poisoning is usually caused by the *sugar of lead*, which is to be neutralized by sulphuric acid in some form, or the soluble sulphates. A teaspoonful of dilute sulphuric or aromatic sulphuric acid and an ounce of Epsom salts in water may be administered as an antidote. Vomiting is to be produced, and bland drinks should be administered, together with a purgative to clear the bowels.

Copper.—Poisoning by copper salts is to be treated by producing vomiting and administering flour water, white of eggs, and milk, in large quantities, both as antidotes and demulcents.

Iodine in the form of the tincture, taken internally, is to be neutralized by starch and water.

Irritant vegetable and animal substances.— In poisoning by these substances, which include *croton-oil, essential oils, cantharides*, etc., free vomiting should be induced, and demulcent drinks given in copious quantities.

Poisoning by **tainted meat, tainted fish, or toadstools,** etc., is to be treated by inducing free vomiting, administering demulcent drinks, and giving a purgative to clear the bowels.

Neurotic Poisons manifest their action mainly through the nervous system; some of them, however, have a local irritant action in addition. The former are narcotics, simply; the latter, irritant *narcotics*. The best known among narcotics are opium, chloral, and hydrocyanic (prussic) acid, but aconite, nux vomica, belladonna, digitalis, tobacco, etc., may also be cited as irritant narcotics.

In connection with vegetable poisons of the neurotic class, it should be remembered that the particular substances which cause poisoning are such as form comparatively harmless compounds with tannin: hence the efficacy of strong hot tea or coffee in these cases, not only as a stimulant, but as an antidote, should be borne in mind.

Opium.—Opium is a *narcotic*. Opium-poisoning may be caused either by the solid opium, morphine, laudanum, or the thousand and one painkilling preparations of the drug.

When poisoning has occurred the stomach is to be promptly and repeatedly emptied by an active emetic (mustard, assisted in its action by tickling the throat with a feather). Tannin and strong hot tea is then to be liberally given. The narcotic effect of the drug must be vigorously combated. The patient should be roused by flapping him with a wet towel, spanking him with a brush or slipper, walking him about, pinching him, shouting at him, etc. Ammonia may be applied to the nostrils, and a pint of hot tea or coffee injected into the bowels. His head may be frequently douched by pouring cold water from a height, the patient being dried off at intervals. Artificial respiration should be employed when the breathing falls below eight a minute; it may be continued for hours. Warmth is to be applied if the bodily temperature fails.

Chloral calls for much the same treatment as opium.

Hydrocyanic Acid, or *prussic acid* is so rapid in its action and so readily absorbed, that there is no time for the employment of emetics. The bearer will have to confine himself to counteracting the paralyzing effects of the drug by cold douches, applications of ammonia to the nostrils, stimulant injections into the bowels, and artificial respiration.

Irritant Narcotics.—In poisoning from *aconite, belladonna, atropine, nux vomica, strychnine, digitalis, tobacco, hemlock, Jamestown weed,* etc., the stomach is to be promptly and repeatedly emptied by an emetic. Tannin, hot tea, coffee, and stimulants are then to be administered, warmth applied, if necessary, and artificial respiration practised, if called for. The powerful antidotes are not safe in the hands of the bearer, and should be administered only by the surgeon.

By **Poisoned Wounds** are meant such as have been poisoned by the instrument or agent inflicting them. They include the bites inflicted by certain animals and insects, dissection wounds, etc.

The bite of a venomous serpent is to be treated by preventing absorption of the poison, destroying or removing it, and combating symptoms of collapse. Absorption is prevented by tying a cord tightly around the limb immediately above the wound. The poison may be removed by forcibly sucking the wound, provided the lips are not chapped or sore; it may be destroyed by burning the wound with gunpowder poured in it and touched off by a live coal, etc., or by cauterizing with ammonia, caustic, acid. To combat the constitutional symptoms of the poison, rest should be secured and alcoholic stimulants given in large doses until intoxication is produced, which should be kept up until arrival of the surgeon.

The bite of a mad dog or other rabid animal requires treatment similar in principle to that detailed above. Alcoholic stimulants, however, are not necessary.

The stings of tarantulas, centipedes, insects, etc., are to be treated with cold applications. Hartshorn applied to the part will neutralize the poison.

Rhus Poisoning.—Poison-ivy, poison-oak, and poison-sumach often produce an eruption of the skin attended with a good deal of redness, itching, swelling, and even blistering.

Weak alkaline solutions, lime-water, applied on lint, are useful. Laudanum may be used to relieve pain.

TABLE OF POISONS.

NAME.	FIRST-AID TREATMENT.
UNKNOWN	Induce vomiting repeatedly, Give bland and soothing drinks, Relieve pain, Stimulate in case of collapse, Employ artificial respiration if necessary.

CORROSIVES.

Acids: Sulphuric (oil of vitriol), Nitric, Hydrochloric (muriatic)	Give alkaline antidotes, Induce vomiting, Give bland liquids, Stimulate if necessary, Secure rest, Relieve pain.
Oxalic	Give mixture of water and chalk, magnesia, lime, or whitewash, or lime-water, *Give no alkalies*, Induce vomiting, Give soothing drinks, Stimulate, if necessary, Secure rest, Relieve pain,

Name.	First Aid Treatment.
Corrosives.	
Alkalies: Ammonia, Soda, Potash, Lye, Hartshorn liniment,	Give acid antidotes, especially vinegar, Induce vomiting, Give bland liquids (e.g., sweet oil), Secure rest, Stimulate, if necessary, Relieve pain.
Corrosive sublimate	Give whites of egg (raw), or tannin (solution, strong tea, coffee), Induce vomiting, Give eggs and milk, Give a dose of castor-oil, Secure rest, Relieve pain.
Nitrate of silver (lunar caustic).	Give solution of salt in large quantities, as antidote and emetic, Give soothing drinks.
Phosphorus	Give calcined magnesia in water, Induce vomiting, Give bland liquids (*no oils or fats*).
Irritant Poisons.	
Arsenic, Paris green.	Induce vomiting at once, Give the "antidote for arsenical poisoning" on hand in dispensary, Dialyzed iron, or Calcined magnesia, Induce vomiting after each dose of the antidote, Give a dose of castor-oil, Soothing drinks, Stimulate, if necessary, Relieve pain, Secure rest.
Tartar emetic	Give tannin (as solution, strong tea, coffee), Encourage vomiting, by large draughts of tepid water, or bland liquids, Give soothing drinks, A dose of castor oil, and Stimulants, if necessary.

NAME.	FIRST AID TREATMENT.
IRRITANT POISONS.	
Lead preparations (sugar of lead, etc.)	Give Epsom salts, in water, Induce vomiting, Give soothing drinks, Dose of castor-oil.
Copper preparations	Induce vomiting, Give flour and water, White of eggs (raw), or Milk in copious quantities, Dose of castor-oil.
Iodine	Give starch and water, Induce vomiting.
Essential oils, Croton oil, Cantharides (Spanish fly).	Induce free vomiting, Give bland drinks in copious quantities, Relieve pain.
Tainted meat, Tainted fish, Toadstools.	Induce free vomiting, Give bland liquids, Castor-oil (a dose).
NEUROTIC POISONS.	
Narcotics: Opium, Laudanum, Morphine.	Induce vomiting by giving mustard-water and tickling throat, Give tannin (solution, strong tea, coffee), Keep patient aroused, Employ artificial respiration, Apply warmth, Stimulate.
Chloral	Induce vomiting, Keep patient aroused, Employ artificial respiration, Apply warmth, Stimulate.
Hydrocyanic acid (prussic acid), Cyanides.	Apply ammonia to nostril, Employ stimulant injections into bowel, Artificial respiration.
IRRITANT NARCOTICS.	
Aconite (monk's-hood), Belladonna (nightshade), Atropine, Nux vomica, Strychnine, Hemlock, Jamestown weed, etc.	Induce vomiting, Give tannin (solution, hot tea, coffee), Stimulants, Apply warmth to body, Artificial respiration, Secure rest, Relieve pain.

NAME.	FIRST AID TREATMENT.
Poisoned Wounds.	
Bites of venomous serpents...	Tie cord tightly around limb above wound, Cauterize wound with acid, ammonia, live coal, Touch off gunpowder in wound, Suck wound if lips are not chapped or teeth hollow, Cut out the part bitten, Produce intoxication.
Bites of rabid animals........	Treat as above, with the exception of producing intoxication.
Stings of tarantulas, centipedes, insects.	Apply ammonia, Employ cold applications, Stimulate, if depression is caused.

FOREIGN BODIES IN THE EYE, EAR, OR NOSE.

Foreign Bodies in the Eye, such as cinders, sand, chips of metal, etc., may be removed in various ways. Frequently the eye relieves itself of the irritating substance by washing it out with a copious flow of tears.

Should this natural method of dislodgement fail, and the foreign body is seen on the globe of the eye, it may be brushed away with a camel's-hair pencil or the corner of a handkerchief; if partly imbedded, it should be lifted out with the point of a lancet or scalpel, the instrument being applied on the flat and used with extreme caution.

Again, the upper eyelid may be seized by the lashes and drawn down over the lower one. When it is released, its under surface is swept by the lower eyelashes, and if the foreign body is within reach, it will probably be caught, as on a brush.

Another method of removing a foreign body from the under surface of the upper eyelid is as follows:

The upper eyelid being drawn down, a probe, slender pencil, toothpick, or match is placed firmly against it, parallel to its edge, and close to the margin of the orbit. The upper lashes are then seized by the disengaged hand, the patient is instructed to look downwards, and the lid is then gently folded backward over the probe or its substitute. The foreign body is then to be brushed away with a camel's-hair pencil, or the corner of a handkerchief, or, if imbedded, carefully removed with the flat surface of a blunt-edged scalpel, lancet, or pen-knife.

The under surface of the lower eyelid is very readily examined.

After removal of a foreign body the irritation caused by it may be soothed by instilling a few drops of olive- or castor-oil, and, if necessary, by cold applications.

It must be borne in mind that often after removal of a foreign body a sensation remains as if it were still present.

When **caustic substances** like lime, ammonia, and acids have entered the eye, the organ should be promptly deluged with water. Lime and ammonia are then to be neutralized by diluted vinegar or lemon-juice (teaspoonful to teacupful of water), and acids by weak solutions of bicarbonate of soda.

Foreign Bodies in the Ear.—Insects in the ear may be dislodged by holding the head with the affected ear upward and filling the canal with warm oil, glycerine, or salt and water. The insect will in short time drown and float to the surface of the fluid, when it may easily be removed.

Wax is to be softened by oil dropped into the canal, which is then to be plugged with a pledget of cotton; after a number of hours the pledget may be removed, and the canal syringed out with warm water until the wax is discharged.

Bodies within easy reach may be removed by the forceps or a hair-pin.

If the foreign body is not accessible, the ear should be carefully syringed.

In **syringing the ear,** the method to be pursued is as follows:

The patient or an assistant holds a cup or vessel under the ear to catch the fluid. The bearer seizes the lobe of the ear and draws it upward, outward, and backward, in order to straighten out the canal. The point of the ear syringe is directed successively to every side of the foreign body, care being taken not to use too much force in delivering the stream of warm water.

Instead of the ear syringe a Davidson or fountain syringe may be used, the head of the patient being held face downward. The nozzle of the syringe, however, is not to be introduced into the canal, as it might prevent the escape of the foreign body.

Foreign Bodies in the Nose.—In cases of foreign body in the nostril, the bearer should close the affected nostril with the finger and instruct the patient (generally a child) to take a deep breath. Then removing his finger, closing the other nostril with it, and the mouth with his hand, he causes the patient to exhale forcibly, clapping him smartly on the back as he does so. By this procedure the foreign body may be dislodged.

If the object is near the opening of the nostril, the bearer may, with his fingers, compress the nose above it and effect removal by means of a pair of forceps, hair-pin, toothpick, bent wire.

The surgeon's services should be obtained without delay if the above measures prove unsuccessful.

MISCELLANEOUS HINTS.

Constipation.—A Seidlitz powder taken before breakfast will relieve a mild constipation. A tablespoonful of Rochelle or Epsom salts before breakfast, a teaspoonful of compound liquorice powder, three or four compound cathartic pills late at night, or a tablespoonful of castor-oil are safe purgatives.

Colic is a violent pain in the bowels caused by the presence of undigested food. A dose of castor-oil will remove the offending material. Hot applications may be applied to the belly if practicable, and a teaspoonful of Squibb's mixture, ginger essence, given in water.

In **Cholera Morbus** the first thing to do generally is to subdue the pain. Half a teaspoonful of laudanum, a quarter of a grain of morphine, or a teaspoonful of Squibb's mixture diluted may be given at once. At the same time hot applications should be applied to the belly if practicable. Stimulants may be necessary in case of collapse.

Diarrhœa is often an effort of nature to rid the bowels of hurtful material. To encourage this effort, a dose of castor-oil may be given. If, after the oil has acted, diarrhœa still persists, a teaspoonful of Squibb's mixture diluted, or a camphor and opium pill, may be administered, to be repeated in an hour or two, if necessary.

THE SIGNS OF DEATH.

The Signs of Death.—Frequently the occurrence of death will be immediately apparent from the character of existing wounds or injuries: should this not be the case, the bearer will look for absence of circulation of the blood, absence of breathing, general ashy paleness of the skin, and its deep discoloration from settling of the blood in those parts which in the position of the body are undermost; dulness and glazing of the eye, cooling of the body; the *rigor mortis*, or death-stiffness coming on, as a rule, after some hours; and finally putrefaction.

Absence of circulation, clearly shown, is a positive sign of death. Cessation of the action of the heart may be established by listening carefully over the region of the heart for its sounds, or feeling for its beat to the left of the breast-bone in the space between the fifth and sixth rib.

Absence of breathing will be apparent from the following: The sound of breathing cannot be heard when the ear is applied to the chest, and no chest motion is perceptible. A mirror or highly polished surface held to the mouth will not become coated with a film of moisture, as it would if held to the mouth of a breathing person. A tuft of down, etc., placed upon the lips will fail to disclose any movement of air-currents.

The bearer will do well not to rely upon any limited number of signs. The above are the principal ones; putrefaction is not necessary to establish death when they are associated with a number of the others.

BOOKS OF REFERENCE.

Encyclopædic Index of Medicine and Surgery.   Bermingham.
Reference Handbook of Medical Sciences.   Buck.
Physiology.   Dalton.
What to do First in Emergencies.   Dulles.
Manual of Drill for the Hospital Corps and Company Bearers.   Havard.
Instruktionsbuch für den Krankenträger.   Hering.
The Surgeon's Handbook.   Porter.
Handbook for the Hospital Corps.   Smart.
Military Hygiene.   Woodhull.

INDEX.

A.

	PAGE
Abdomen, organs of	15–18
" wounds of	23
" " " position of patient upon litter in cases of	22, 23
Accident, general rules in cases of	46
Acids, poisoning by	64, 65, 69, 71, 72
" in the eye	75
" as antidotes	62
Aconite, poisoning by	70–73
Administration, forcible, of remedies	63
Aid, first, on the battle-field	21–23
Alkalies, poisoning by	65–72
Ammonia, poisoning by	65–72
" in the eye	75
Animation, suspended	43–45
Antidotes, classification of	62, 71–74
" alkaline	62–72
" acid	62–71
Aorta	14–18
Apoplexy	48, 49
Arsenic, poisoning by	66–72
Arteries, the	18–20
" bleeding from (see Hemorrhage)	27
" location of the principal	18–20
Artificial respiration	44
Asphyxia	51–54

	PAGE
Astringents	78
Atropine, poisoning by	70–73
Axis, cerebro-spinal	11

B.

Backbone	5
Bandage, roller	25
" triangular	26
" elastic	26
Bearers, equipment of	24
" limitation of duties of	24
" Company	24
" " duties of	21–23
Belladonna, poisoning by	68–73
Belly, the	15–18
" " wounds of the	23
" " " " " position of patient upon litter in	23–35
Bending of limb to produce pressure upon artery	30, 31
Bile	16
Bile-duct	16
Bite of serpent	70–74
" mad dog or other rabid animal	70, 71–74
Bladder, the	17
Bland liquids	62, 63
Blisters, from marching	57, 58
Blisters, puncture of	58
Bleeding (see Hemorrhage)	58, 59
Blood-vessels, location of principal	18–20
Body, the human	2–20
Bones, the	2–8
Bowels, the	16
" " bleeding from	59
Brain, the	10, 11
" compression of	48
" concussion of	48
Breathing	13
" rate of	13
" artificial (see Respiration, artificial)	44, 45

	PAGE
Bruises	34
Burns	55, 56

C.

Canal, spinal	5
Cantharides, poisoning by	66, 73
Capillaries, the	14
Capsule of joint	8
Carbolic acid, poisoning by	65
Cartilage	6
Caustics in the eye	75
Chest, the	5, 12
" bones of the	5
" organs of the	12
Chilblains	57
Chloral	69
Choking	54
Cholera Morbus	78
Circulation, the general	13–15, 18, 19
" systemic	13, 14, 18, 19
" pulmonary	13–15
Coffee, hot, as an antidote	63, 64, 69, 70, 72, 73
Colic	77
Collapse (see Shock)	33, 37, 38
Column, spinal	5, 11
Compound fractures	38, 39
Compress	25
Compression of the brain	48
Concussion of the brain	48
Connective tissue	9
Consciousness, loss of	47
Constipation	77
Contusions	34
Copper, poisoning by	68, 73
Cord spinal	11
Corps, Hospital	24, 25
Corrosive sublimate, poisoning by	65, 66
Cranium, the	3
" bones of the	3

	PAGE
Cranium, organs of the cavity of the	10, 11
Crepitus	38
Croton-oil, poisoning by	66, 73
Cuts	15, 23, 27, 35
Cut belly, position of patient in transportation	35
Cut throat, position in which head should be kept in	23

D.

Death, signs of	78, 79
Diaphragm	12
Diarrhœa	18
Digitalis, poisoning by	68, 73
Dislocation	43
Dog, bite of mad	70, 74
Dressing station	22
Drowning	51–53
Drunkenness	50

E.

Ear, foreign bodies in the	76, 77
Emetics applicable in cases of poisoning	61, 62
Epileptic fit	50
Equipment, the bearer's	24
Esmarch's package	24
Essential oils, poisoning by	68, 73
Excretions	17, 54
Exhaustion from heat	49, 50
Eye, foreign bodies in the	74, 75

F.

Face, the	4
" bones of	4
Fainting	33
Feet, soreness of the	57
Fish, tainted, poisoning by	68
Fingers, used to apply pressure	7, 28–31
Fit, epileptic	50
Flexion of limb to produce pressure upon artery	30, 31

	PAGE
Foreign bodies in the eye	74, 75
" " " ear	76, 77
" " " nose	77
" " " windpipe and gullet	54
Foxglove, poisoning by	70, 73
Fractures in general	38
" special	39–42
" compound	38, 42
" position of patient upon litter in cases of	39
Freezing	56
Frost-bites	56, 57

G.

Gall-bladder	16
Gases, suffocation caused by	53
Gastric juice	16
Glands	17
Gristle	6
Gullet	16

H.

Handkerchief to produce pressure in hemorrhage	29
Hanging, strangulation from	53
Head, organs of the lesser cavities of the	12
Heat exhaustion	49, 50
Heart, the	13–15
Hemlock, poisoning by	70, 73
Hemorrhage, capillary	27, 31–33
" in general	26–30
" venous	28
" arterial	28–30
" from the artery of the neck	30
" " " " " arm in its uppermost portion	30
Hemorrhage from the artery of the arm in its lower portion	31
Hemorrhage from the arteries of the fore-arm	31

	PAGE
Hemorrhage from the thigh	32
" " " leg	32
" " " foot	32
" internal	32
" on reaction from fainting	33
" from the nose	58
" " " lungs	58, 59
" " " stomach	59
" " " bowels	59
" " " great cavities, position of patient upon litter in cases of	22, 35
Hydrochloric acid, poisoning by	65, 71
Hydrocyanic acid, poisoning by	68, 73

I.

Insects, in ear	76
" sting of	71, 74
Insensibility	48, 49
Intestines	35
Intoxication	50
Ipecac as an emetic	62

J.

Jamestown weed, poisoning by	70, 73
Joints, the	8
" parts forming	8
Juice, gastric	16

K.

Kidneys, the	17
Knotted cloth used to apply pressure in hemorrhage	29

L.

Laudanum, use of, in case of pain from poisoning	63, 71
" poisoning by	73
Lead, poisoning by	67

INDEX.

	PAGE
Ligaments, the	8
Limbs, upper, bones of	6
" lower, bones of	6
Lime in the eye	75
Lint and its substitutes	25, 35
Liquids, bland	62, 63
Litter, position of wounded upon	22, 23
Liver, the	16
Lungs, the	12, 13
" bleeding from	58, 59
" position of patient upon litter in cases of wound of,	23
Lye, poisoning by	65, 72
Lymphatics, the	10

M.

Meat, tainted, poisoning by	68
Medicine chest, the bearer's	24
Membranes of brain	10
Method, Silvester's	44, 45
Midriff, the	12
Morphine, poisoning by	68, 73
Muriatic (hydrochloric) acid, poisoning by	64, 71
Muscles, the	9

N.

Neck, position in which head should be kept in wounds of front of	23
Nerves	10–12
Nitrate of silver, poisoning by	64, 66, 72
Nitric acid, poisoning by	71
Nose, bleeding from	58
" foreign body in	77

O.

Oil, Croton, poisoning by	68, 73
Oils, essential, poisoning by	68, 73

	PAGE
Oil of vitriol (sulphuric acid), poisoning by	65, 71
Opium, poisoning by	69, 73
Oxalic acid, poisoning by	65, 71

P.

Package, Esmarch's	24
Pad and bandage, to produce pressure in hemorrhage	29, 30
Pancreas, the	17
Paris green, poisoning by	66, 72
Parts, the soft	9
Pelvic cavity, organs of	15–17
Pelvis, bones of	6
Peritoneum, the	18
Phosphorus, poisoning by	66
Pleuræ, the	12
Plug and bandage to produce pressure in hemorrhage	28
Poisoning	59–74
" definition of a poison	59
" classification of poisons	59
" general measures and remedies	60, 61
" by unknown substances	64
" by corrosive substances	64
" sulphuric, nitric, and hydrochloric acids	64
" carbolic acid	65
" oxalic acid	65
" ammonia, potash, and soda	65
" corrosive sublimate	65, 66
" lunar caustic, nitrate of silver	66
" phosphorus	66
" by irritant metallic substances	66
" arsenic	66, 72
" tartar emetic	67, 72
" lead	67, 73
" copper	68, 73
" by irritant vegetable and animal substances	68, 73
" by tainted meat, tainted fish, toadstools	68, 73
" by narcotics	68, 73
" opium	69, 73

INDEX.

PAGE

Poisoning by chloral..................................69, 73
" hydrocyanic or prussic acid......... ....69, 73
" by irritant narcotics.................. ....70, 73
" aconite, belladonna, digitalis............70, 73
" Jamestown weed, hemlock, etc...........70, 73
Poison ivy and poison-oak eruption....... 71
Poisoned wounds......................................70, 74
Potash, poisoning by..................................65, 72
Powder for sore feet............................ ....57, 58
Pressure, methods of applying, in hemorrhage.........28–33
Prussic (hydrocyanic) acid, poisoning by................69, 73

R.

Respiration, rate of..................................... 13
" artificial.................................. 44
" Sylvester's method of artificial...........44, 45
" (see Breathing)........................... 13
Rhus poisoning.. 71

S.

Salt as an emetic...................................... 61
Scalds... 56
Secretions... 17
Seizures, epileptic..................................... 50
Seminal vesicles....................................... 17
Serum.. 18
Shock accompanying severe injury..................... 37
" fright, despondency, hunger, and thirst as favoring, 37, 38
" as a result of extensive burns or scalds............ 56
Silver, nitrate of, poisoning by......................66, 72
Silvester's method....................................44, 45
Skeleton, the..................................... 2, 3
Skin, the..9, 10
Skull, the....... ....3, 4, 10
Snake-bites..70, 74
Soda, poisoning by..................................65, 72
Soft parts, the..9–11

	PAGE
Soreness of the feet	57
Spine	5, 11
Spinal column	5
" cavity, organs of	10
Spleen, the	17
Splints	26
" improvised	26, 39–42
Sprains	42
Stimulants in cases of poisoning	63, 66, 68–73
Stings, of tarantulas, centipedes, insects, etc.	71, 74
Strangulation from hanging	53
Stomach, the	16
" bleeding from	59
Strychnine, poisoning by	68, 73
Stunning (see Concussion of the Brain)	48
Styptics	25, 27, 29
Sublimate, corrosive, poisoning by	64, 72
Suffocation with gases	53
" from foreign bodies in windpipe or gullet	54
Sugar of lead, poisoning by	67, 73
Sulphate of zinc, as an emetic	62
Sulphuric acid, poisoning by	64, 65, 71
Sunstroke	49
Sutures of the skull	4

T.

Tannin as an antidote	67, 72, 73
Tartar emetic, poisoning by	67, 72
Tea, hot, as an antidote	68, 69, 72, 73
Teeth, the	4
Temperature, normal, of the body	15
Thorax, the	5
" bones of the	5, 6
" organs of the cavity of the	12–15
Tickling the throat to produce vomiting	61, 69, 73
Tourniquet	26, 29–32, 37
Tobacco, poisoning by	70, 73

U.

	PAGE
Unconsciousness	47
" position of patient upon litter in cases of,	47, 48
Ureters	17
Urethra	17
Urine, the	17
" daily quantity of	17

V.

Veins, the	14, 20
" location of principal	14, 18, 19, 20
" bleeding from	15, 27, 28
Vomiting, how produced	61, 62, 66, 69
" when to be produced	64–74

W.

Warm water as an emetic	61, 62, 72
Wax in ear	76
Windlass, Spanish	26, 29–32
Wounds in general	21, 27, 33
" contused	34
" incised	35
" punctured	35
" lacerated	36
" gunshot	36
" poisoned	70
" of belly	23, 35
" position of patient upon litter in cases of	23, 35
Wounded in battle, general management of	21–23

Z.

Zinc sulphate as an emetic	62

www.ingramcontent.com/pod-product-compliance
Lightning Source LLC
Chambersburg PA
CBHW030052170426
43197CB00010B/1496